Studies in History and Philosophy of Science

Volume 57

Studies in History and Philosophy of Science is a peer-reviewed book series, dedicated to the history of science and historically informed philosophy of science. The series publishes original scholarship in various related areas, including new directions in epistemology and the history of knowledge within global and colonial contexts. It includes monographs, edited collections, and translations of primary sources in the English language. These cover a broad temporal spectrum, from antiquity to modernity, and all regions of the world.

Harald A. Mieg

Editor

The Responsibility
of Science

EUROPEAN COOPERATION
IN SCIENCE & TECHNOLOGY

Springer

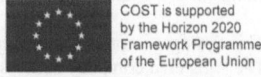

COST is supported
by the Horizon 2020
Framework Programme
of the European Union

Editor
Harald A. Mieg
Institute of Geography
Humboldt-Universität zu Berlin
Berlin, Germany

This book was written in the broader context of RECIPES "Precaution, Innovation, Science." The RECIPES project has received funding from the European Union's Horizon 2020 research and innovation programme under grant agreement No. 824665.

This publication is based upon work from COST Action CA18204, supported by COST (European Cooperation in Science and Technology).
COST (European Cooperation in Science and Technology) is a funding agency for research and innovation networks. Our Actions help connect research initiatives across Europe and enable scientists to grow their ideas by sharing them with their peers. This boosts their research, career and innovation.
www.cost.eu

Proofreading: Dave Morris

ISSN 0929-6425   ISSN 2215-1958   (electronic)
Studies in History and Philosophy of Science
ISBN 978-3-030-91596-4   ISBN 978-3-030-91597-1   (eBook)
https://doi.org/10.1007/978-3-030-91597-1

This Springer imprint is published by the registered company Springer Nature Switzerland AG
The registered company address is: Gewerbestrasse 11, 6330 Cham, Switzerland

Preface

Our volume on the responsibility of science stems from a 2019 meeting of GeWiF, the Berlin Society for the Study of Science (Gesellschaft für Wissenschaftsforschung). The opening question of the meeting also defines the topical outline of this book: *Science has become indispensable. It has long since left behind the stage of pure observation and the acquisition of knowledge. Instead, in the twentieth century, science became a power through its alliance with technology. The development of nuclear energy, genetic engineering, or digitalization is no longer conceivable without the involvement of science. This power gives rise to responsibility. How far [then] does the responsibility of science actually extend? How far does it go beyond the obligation to conduct expert research?*

GeWiF was founded in 1991 and can be traced back to Albert Einstein's initiative to establish a professorship on the philosophy of physics for Hans Reichenbach at Berlin University (then Friedrich-Wilhelms-Universität, now Humboldt-Universität zu Berlin). This succeeded in 1926, and in 1928, with Reichenbach on the board, the Society for Empirical Philosophy (Gesellschaft für empirische Philosophie) was founded. In 1930, Reichenbach, together with Rudolf Carnap, took over the publication of the journal *Erkenntnis* (Knowledge). With the Nazi's ascent to power in 1933, Reichenbach was dismissed. He then emigrated, later settling in the United States in 1938, where he taught at the University of California at Los Angeles (UCLA) until his death in 1953. *Erkenntnis* was re-established in 1979 by the scientific philosophy community. In the first issue from 1930, Reichenbach had written that he wished to continue the tradition of "doing philosophy not as an isolated science, but in the closest connection with the individual disciplines."[1] This was exactly what the philosopher Heinrich Parthey subsequently pursued in founding and leading GeWiF for many years. The Society's conferences involved researchers in economics, sociology, and industry, but also in state administration. Recurring themes included innovation through science as well as the ongoing institutionalization of science.

---

[1] Reichenbach, H. (1930). Zur Einführung. *Erkenntnis* 1(1), 1930, p. 1. (translated)

This context determines the particular focus and publication mode of our volume, which can be characterized in three ways. First, the Germanic context, which arises mainly from the members of the GeWiF, some of them having experienced the immense social shifts from the Nazi regime to the former East Germany (German Democratic Republic, 1949–1990) under the influence of Soviet ideology, and, finally, today's Germany, which understands itself as an open democratic society. This biographical impact also became evident in other topics that GeWiF addressed in its meetings, such as science as a profession or the role of critical thinking in science and society. The second characteristic is the embedding into European research. Topics and funds for research in Europe are essentially determined by European Union research planning, with the EU level also vigorously addressing the responsibility of science. GeWiF decided therefore to publish the present volume in English. The publication would not have been possible without the support of two EU research projects: RECIPES ("Precaution, innovation, science," https://recipes-project.eu/) and the COST Action CA18204 ("Dynamics of placemaking and digitization in Europe's cities," www.placemakingdynamics.eu/). Thirdly, understood as an aspect of its own scientific responsibility, GeWiF has always published in open access.

We dedicate this book to Heinrich Parthey, the founder and long-term president of GeWiF, who died in 2020 at the age of 83.

Berlin, Germany   Harald A. Mieg
March 2022   Hans Lenk
                                                                         Rainer E. Zimmermann

Contents

Chapter 1
The Responsibility of Science: An Introduction

Harald A. Mieg

Abstract This is the introduction to the book *The Responsibility of Science*, containing three parts. I explain both the concept of responsibility and science as an institution. I then present lines of argumentation that run through the essays of this volume and combine them. (i) Responsibility is a relational concept, derived from the verb "to respond." Therefore, the concept of responsibility refers to a relation involving at least three elements: Someone is responsible for something to someone else. Moreover, responsibility is attributive, that is, resulting from a social attribution of guilt or duties to a person. (ii) Science is meant here to refer to historically developed, institutionalized research and to be thought of independently of the objects of that research. Therefore, by 'science,' I am referring to natural and social sciences as well as humanities, and make no distinction between pure and applied science. (iii) This volume lives through the many references that link the chapters and the lines of argumentation that develop in the work, such as Responsible Research and Innovation (RRI) as a new approach within EU research policy; the ethical question of the moral person in science; and the effects of the institutionalization and professionalization of science.

Immediately prior to his death, Einstein was drafting a public speech intended to mediate in the Israeli–Egyptian conflict. Indicating that he was addressing issues of personal responsibility primarily as a member of humankind, it began: "In matters concerning truth and justice there can be no distinction between big problems and small; for the general principles which determine the conduct of men are indivisible."[1] He died before finishing the draft, silencing his passionate voice for peace. Einstein's note might suggest a parallelism of scientific and ethical that would be disputed by those in science. Questions of truth are usually settled differently than questions of

---

[1] O. Nathan & H. Norden H. (ed.), Einstein on peace. Schocken books, NY. p. 639.

H. A. Mieg (✉)
Institute of Geography, Humboldt-Universität zu Berlin, Berlin, Germany
e-mail: harald.mieg@hu-berlin.de

© The Author(s) 2022
H. A. Mieg (ed.), *The Responsibility of Science*, Studies in History and
Philosophy of Science 57, https://doi.org/10.1007/978-3-030-91597-1_1

justice. Science feels first and foremost committed to the search for truth and valid evidence. Do ethics disturb science, we may ask, or is science in some way based on ethics? Do we find any justification for ethical issues in the scientific task, or in the people who do science?

Much has been written about the ethics of science since the first atomic bomb was dropped. I would not dare to publish this book if I felt that today the question of the responsibility of science would mainly refer to problems of the type "Should nuclear research be used for bombs?" This type of problem challenges the scientist as a moral person who must come to terms with their conscience. The current issue of climate change has a different quality, as knowledge is being driven forward by the scientific community through public research communication together with massive of deployment of human and financial resources. This type of project would be inconceivable for some eminent scientists, who repeatedly admonish politics. Rather, climate change research is evidence of a profound institutionalization of science and is supported by new scientific organizations such as the IPCC (Intergovernmental Panel on Climate Change). I am convinced that the current forms of scientific institutionalization are an expression of its professionalization. Science has become a *profession*. This changed the focus from ethics to social responsibility, which is not a big change and neither makes the ethical question disappear nor provides new answers concerning the social contribution of science. The difference is that today, in addition to an individual responsibility, there is a *corporate* responsibility of science—as a profession that may have to prove its social responsibility as do some other professions. Since science as a profession only emerged in the twentieth century and is thus young compared to doctors or engineers, the question of the responsibility of science must also be asked anew, for it cannot be so easily reduced to the responsibility of individual scientists. Hence the subtitle of our book: essays on the extents of the scientific profession's (moral) responses to societal concerns.

This volume has three parts. The first part is entitled "Principles of the Responsibility of Science" and discusses the concept of responsibility, not least with a focus on corporate responsibility. The second part has the awkward title "Insights into the quest for responsibility in the interaction of science and society". If science is a profession and enjoys not only freedoms but also considerable resources, then it must be able to justify their benefits and in this context, even more than before, keep in mind the sometimes-unintended consequences of scientific work. This second part demonstrates how research policy and legislation can respond, for example in the paradigm of RRI (Responsible Research and Innovation) or the Precautionary Principle. The third part unites reflections on the topic of "science and responsibility," ranging from a very personal, emotional statement to an institutional program. This third part testifies both to the effort to negotiate the scientific discourse for truth against social appropriation and relativization, as well as new forms of value-based coproduction of knowledge of science and society.

Responsibility

Responsibility is a *relational* concept, derived from the verb "to respond."[2] Therefore, the concept of responsibility refers to a relation involving at least three elements: Someone is responsible for something to someone else. If the context is clear, we can use abbreviated versions such as "someone is responsible for something" or simply "someone is responsible."

Moreover, responsibility is *attributive*, that is, resulting from a social attribution of guilt or duties to a person. Responsibility is often classified according to the type of attribution: moral responsibility may be evoked in case to attributing guilt; role responsibility in the context of allocating socially defined tasks to someone; social responsibility is linked to the attribution of unspecified prosocial contribution, etc.

Further characteristics are:

- To some extent, responsibility presupposes both *causality* and *freedom*. Responsibility requires causality; For if we could not cause anything through our actions, we could not be held responsible. Responsibility also requires freedom; For only if we have freedom to act in a chosen way can we be held responsible. Total causality (everything is predetermined) as well as total arbitrariness (what happens is mere chance) are incompatible with the idea of responsibility.
- Responsibility has an inherent *temporal* component. Because either something has already happened and the person responsible is being sought, or someone is given the responsibility to take care of something. Therefore, we also speak of retrospective and prospective responsibility.
- Insofar as responsibility is socially relevant, it concerns behavior among people and is therefore *morally* relevant. This has ethical implications.

Responsibility has much in common with the concept of accountability. The difference is the inclusion of the temporal dimension, so that we can explicitly speak of a future responsibility.

With the question of responsibility, we enter the realm of ethics, in which, however, the concept of responsibility has long played a subordinate role. Ethics tends toward absoluteness, as in Kant's categorical imperative. Categorical, absolute principles such as "Thou shalt not lie" are useful as ideals and regulative ideas. Taken absolutely, however, they are almost useless, for example in both politics and everyday life. For this reason alone, Max Weber called for responsibility to be made a principle, in marked contrast to absolute ethics.[3] In philosophy, the concept of responsibility was disregarded for a long time, because it seemed to raise more questions than it answered. It was not until the publication of *Das Prinzip Verantwortung* (1979; English: The Imperative of Responsibility, 1984) by Hans

---

[2] For details, please refer to the chapters in Part 1.

[3] Weber, M. (2004). Politics as a vocation. In M. Weber, The vocation lectures (edited by D. Owen & T. B. Strong, translated by R Livingstone, pp. 32–94). Indianapolis: Hackett. (The lecture took place in January 1919).

Jonas that responsibility received new attention. Jonas focused on our responsibility for the future: that we maintain a grip on what is technically possible and do not unintentionally destroy our own basis of life. Science has very much its own ambivalent role in this, on the one hand in supporting innovation and prosperity and finding ways to improve our lives; on the other hand, in developing technologies that we humans no longer seem able to control. Hence, with such far-reaching capabilities of science come myriad forms of responsibilities.

Science

Science is meant here to refer to historically developed, institutionalized research and to be thought of independently of the objects of that research. Therefore, by 'science,' I am referring to natural and social sciences as well as humanities. I also make no distinction here between pure and applied science. Since science has existed in the form of systematic research, it has always served many purposes, new methods of growing crops, as well as healing and warfare.

Science has characteristics that can quickly lead to ethical implications. Three of these characteristics are:

- *Ambivalence*: As such, scientific findings do not necessarily determine the nature of their use; hence ambivalence is inherent in science. In addition to peaceful use, there are sometimes military applications or potential for criminal misuse. Today, this is regarded as a problem of *dual use*. Professionalization has led to dual use being addressed within the framework of codes of conduct.
- *Innovation*: Science is innovative per se, because the goal is new knowledge, including new processes and techniques. In recent decades, science-based innovation in food, medicines, or agricultural technology has sometimes had unintended consequences for the environment and health. Therefore, policy and legislation must find new ways of risk assessment and precaution.
- *Formalization*: The mathematical formulation of physical relationships as well as the representation of contractual forms in the language of jurisprudence are examples of formalization in science, which can give outsiders the feeling that their own life world is only reflected in a very reduced way in the formulas. For this reason, some scholars consider a parallelization of machines and humans, as in the paradigm of artificial intelligence, as a dangerous ethical reduction.

For half a millennium, science has been associated with progress. With professionalization, science has gained a new standing in society. Not least because of its unbelievable expansion since the Second World War, science is now concerned with all areas of life and has been able to prove even that general expenditure on research and development (including in science) goes hand in hand with economic growth. Furthermore, science has gained methodological certainty, which does not commit it to the position of value-free research, but also allows it to work scientifally on value-laden issues such as sustainable development. It is becoming increasingly clear that science has more to offer society than simple truths (which are often not so simple), but also an attitude—its own ethos—based, for example, on principles such as transparency and open access.

This present volume is strongly influenced by German philosophy and history. In the nineteenth century, Germany was booming, just as China is today. One driver was science, which was simultaneously a matter for both academia and industry. In 1912, the Kaiser Wilhelm Society (KWG, Kaiser-Wilhelm-Gesellschaft) was founded, the predecessor of today's Max Planck Society for the Advancement of Science. The principle of the KWG was the greatest possible freedom of research, combined with generous endowment, in a socially relevant field—be that theoretical physics or industrial iron production. The benefit was seen equally for government, science, and industry. The use of scientific advancement was thus ambivalent and multi-oriented from the outset. Thus, the use of poison gas in the First World War also fell within the context of KWG research. Responsibility therefore—as opposed to ethics understood in absolute terms—can also have a historical side and may require learning.

Some readers may wonder why medicine does not play a central role in this volume on responsibility of science. The simple answer is that medicine—like law—has been a profession in its own right for centuries. Many of the problems and solutions mentioned in our book have already been played out in the medical profession. The Hippocratic oath is over a thousand years older than the code of codes for scientists. That science is now also a profession will not add much to the ethical questions encountered in medicine. However, new light will also be shed on the large field of associated research, which has long been able to hide behind medical practice. Moreover, in the current global COVID-19 pandemic, the question of the practical responsibility of medical science is highly topical. For many readers, medicine may be considered the most obvious field through which to discuss the responsibility of science. However, the question then immediately arises of whether this can also apply to other fields such as physics and urban planning. Therefore, we have chosen to use examples directly from those other fields.

Lines of Argumentation

Although I have given all chapter in this book a place in the overall argument, each can stand on its own. Readers may begin with any chapter. Apart from the overall argument, this volume lives through the many references that link the chapters and the lines of argumentation that develop in the work. I would therefore like to highlight the most important ones here.

1. *Responsible Research and Innovation* (RRI). RRI is a new approach within EU research policy. Macnaghten (Chap. 5) introduces RRI on the basis of four dimensions: anticipation, inclusion, reflexivity, and responsiveness. In the context of a COST-EU project (in the field of urban planning) involving over 30

countries, the opportunity arose to reflect on one's own ongoing research in the light of RRI (Chap. 12). Macnaghten presents RRI as a further development of the co-production approach, which considers the involvement of citizens as essential ("inclusion"). Oevermann et al. show in Chap. 12 that implementing this requirement for inclusion is anything but easy, even in the field of planning. Nevertheless, surprisingly, on the one hand is the productive contribution of anticipation, e.g., thanks to scenario-building (speaking for RRI?); and, on the other hand, a rather low level of reflexivity in the individual projects (speaking against RRI?), which may also be due to the fact that science today is professionalized throughout Europe. Research, even if imagination remains an important factor, is in practice largely routinized work.

2. *Ethics and the question of the moral person in science.* Hans Lenk's introductory chapter expresses the essence of the moral person (Chap. 2). This serves both as an absolute reference point for ethics (in Lenk's terms "concrete humanity") and as a central point for attribution of responsibility. In Horst Kant's contribution on the history of nuclear fission (Chap. 6), we sense the manifold moral dilemmas of the moral person, which nuclear research imposed upon the various physicists involved. Kant introduces, among others, the renowned physicist and philosopher Carl Friedrich von Weizsäcker, who opposed nuclear armament of West Germany after World War II and was one of the founders of the VDW (Society of German Scientists). The VDW is an association of socially and politically engaged scientists who advocate for the responsibility of science. Their position paper is a prime example of the human-centered attitude with which the VDW approaches the issue of digitization (Chap. 11). Klaus Fuchs-Kittowski makes a similar point, but with considerably more force and emphasis, and his contribution almost resembles a lament or incendiary statement (Chap. 10). In no other chapter does the responsibility of science appear to be so historically anchored and so dependent on the consciences—as moral persons—of every single scientist.

3. *Institutionalization and professionalization of science.* Heinrich Parthey presents the institutionalization of science—via academies, universities, professional societies, national research institutes, etc.—as a guarantor of freedom but also for scientific progress (Chap. 7). According to Parthey, research is always confronted with interdisciplinary research situations, i.e., problems that can only be dealt with by involving different scientific disciplines (horizontal interdisciplinarity) or in cooperation with technical development and industrial production (vertical interdisciplinarity). As my own paper (Chap. 4) discusses, science became almost entirely professionalized during the twentieth century. Science is now a profession just like architecture, with university-supported training programs, professional associations, and career paths that are not always secure. My argument is that this has reorganized responsibilities, creating a new corporate responsibility at the level of self-organized professional institutions. This also means some relief for individual researchers. More people than ever before in history can devote themselves entirely to gaining knowledge on a professional basis without having to justify themselves. Rainer E. Zimmermann's statement

(Chap. 9) can also be seen in this context: The core purpose of scientific discourse must remain to test knowledge for truth.

4. *Precaution and Responsibility.* One of the great interventions, in the sense of Hans Jonas' principle of responsibility, was the introduction of the Precautionary Principle into European legislation. This was the result of the long debates on environmental protection in the late twentieth century. The Precautionary Principle has indirect effects on science, since it usually concerns innovations— for example, nanotechnology, genetically modified organisms (GMOs) in seed production or endocrine disrupting chemicals (EDC) in plastic, paint, toys, clothing, etc.—which affect the material cycles of European societies via industrial production. As de Smedt and Vos explain (in Chap. 8), the introduction of the Precautionary Principle was controversial from the outset. It was feared to have an unfavorable impact on innovation and competitiveness in Europe. To ensure that precaution cannot be reduced to prevention, the idea is to combine the Precautionary Principle with an RRI policy. Formalizing RRI will likely increase the density of regulation and unintentionally reinforce the restrictive effect of RRI policies. It would be worth trying to replace this preventative perspective with an improvement perspective. In environmental management, which corporations sometimes regard as unattractive, the principle of continuous improvement has gained acceptance, and actually become a general principle for progressive process management and refinement. A similar complement to the Precautionary Principle could be the "Principle of Responsive Adjustment" introduced by Peter French (Chap. 3). Responsive adjustment, as a form of continuous improvement, is a maxim that can find appeal not only in industry but also in science, and thus shows the way in which responsibility can be made tangible.

Science today is older than almost all nation-states. The responsibility of science has always been to regulate in dialogue with society. In the past, science depended on the genius and commitment of individual scientists, such as Aristotle or Galileo Galilei. Today, we have powerful scientific institutions and a broad anchoring of science via the integration of universities within society. In return, dialogue has also become more difficult: for example, politics, law, and education are different fields that have developed their own logic. In addition, there is the international dimension. Science has long been globalized. But the impulses are changing. Just as Germany did in the nineteenth century, China is now intervening in the development of science. China is the only country that can be said to be truly older than science. At present, China uses science and innovation for levering national development. China increasingly controls the "boring" but important operational processes of professional science, ranging from academic journals to standardization bodies. If only because of its huge population coupled with economic growth, Chinese investments in biomedical research and digitization are immense. These are two research fields with high innovation and exploitation potential. It is precisely here that the social dialogue on the responsibility of science will have to continue. Probably— and hopefully—science today is a force that, thanks to its institutions, can enter into this dialog about responsibility in a united way where necessary.

Part I
Principles of the Responsibility of Science

This part contains three chapters. The first deals in detail with the concept of responsibility and makes deductions from it for the responsibility of science. The second chapter deals with corporate responsibility. The point we want to make by including this chapter is that science today also assumes corporate responsibility. This gives rise to measures in the political and legal treatment of science. The third chapter explains the responsibility of science as a professional responsibility.

Chapter 2: Hans Lenk's contribution, "Responsibility in Science," is the prelude to this first part, displaying the principles of the responsibility of science. His contribution offers a systematic overview of types of responsibility and introduces prioritization rules. Accordingly, the highest or first priority is given to the principle of concrete humanity. Lenk's chapter concludes with 15 theses on the responsibility of science.

Chapter 3: Peter French's contribution, "The Principle of Responsive Adjustment in Corporate Moral Responsibility," concerns the legal treatment of corporate responsibility. The principle of responsive adjustment, presented to assign responsibility for change, seems appropriate for influencing corporate policy and culture.

Chapter 4: Harald Mieg's contribution, "Science as a Profession, and its Responsibility," assumes that science today is professionalized. With established standards of professionalization, unfortunately, instances of scientific fraud are also to be expected. On the other hand, there are new opportunities to take responsibility at the institutional level, e.g., through the IPCC (Intergovernmental Panel on Climate Change).

About the Authors: *Hans Lenk* (born 1935) contributed the authoritative analyses of responsibility in German-language philosophy. From 2005 to 2008 he was president of the global Institut International de Philosophie. He taught at the Karlsruhe Institute of Technology, Germany. *Peter A. French* (born 1942) is the US counterpart to Hans Lenk. His research has particularly focused attention on corporate responsibility. He taught from 2000 until 2016 at Arizona State University. *Harald A. Mieg* (born 1961) is concerned on the one hand with sociological professionalization and on the other with the philosophy of science. He teaches at Humboldt-Universität zu Berlin.

Chapter 2
Responsibility in Science: The Philosophical View

Hans Lenk

Abstract Terms of responsibility are relational attributes, i.e., attribution terms. They are to be understood as linguistically, socially, and situationally embedded concepts conventionalized by rules and have to be analyzed accordingly. A structural theory of responsibility, and more differentiated forms and types of responsibility such as relational attribution-based concepts, will be developed schematically in order to do justice to the variety of different uses of the concepts of responsibility, e.g., causal and action responsibility, role responsibility, but also social and (universal) moral and legal responsibility. In this chapter, I apply the general considerations of responsibility to analyze responsibility in science. The responsibility of the researcher in science and technology is a special case of role-specific and moral responsibility in a strategic position. Points to be discussed include known means of implementing responsibility in science, including codes of conduct, ethics committees, a scientific ethos, and the Hippocratic Oath for scientists. The chapter concludes with fifteen theses on responsibility in science. The key principle should be "concrete humanity": Practical and concrete humanity should always be a central guiding principle (in dubio pro humanitate practica).

Terms of responsibility are relational attributes, i.e., attribution terms. They are to be understood as linguistically, socially, and situationally embedded concepts conventionalized by rules and have to be analyzed accordingly.[1] A structural theory of responsibility, and more differentiated forms and types of responsibility such as relational attribution-based concepts, will be developed schematically in order to do justice to the variety of different uses of the concepts of responsibility, e.g., causal and action responsibility, role responsibility, but also social, (universal) moral and legal responsibility. These are such complex terms that it is not possible to make an overall general classification. Different types of responsibility would structure the social and normative reality differently and have specific implications. The

H. Lenk (✉)
Karlsruhe Institute of Technology (Emeritus), Karlsruhe, Germany
e-mail: hans@lenknet.de

© The Author(s) 2022

H. A. Mieg (ed.), *The Responsibility of Science*, Studies in History and Philosophy of Science 57, https://doi.org/10.1007/978-3-030-91597-1_2

attribution of responsibility itself can be either descriptive or normative; in such context, it is to be understood as descriptive or normatively acting. Both functions must be carefully (ideally) distinguished in any analysis, even if in practice both attributions are usually made at the same time. Nevertheless, a distinction has to be made between normative and descriptive use(s). Diagrams of, e.g., role and task responsibility as well as moral, legal, and other specific variants, may further subdivide the abstract scheme types or serve for further concretization. The same applies to analytical–structural polarity of responsibilities and to priority rules for handling typical conflicts among some such responsibilities regarding different instances or role-takers. Newer concepts such as social, collective, and corporate responsibility and even system responsibility will require more attention in the future. Even if these analyses are still incomplete, I shall attempt in the following an application to the responsibility in science.

Introduction

In his Dictionary of the Devil (1911) the great satirist Ambrose Bierce defined:

> RESPONSIBILITY: a detachable burden easily shifted to the shoulders of God, Fate, Fortune, Luck, or one's neighbor. In the days of astrology, it was customary to unload it upon a star.

Today, some people would actually rather shift the responsibility to a *star* (in the new societal sense), be that a star of politics, society, or even science. However, although scientists were traditionally considered responsible for "clean" scientific work (today: "good scientific practice") and for successful discoveries, they were not considered responsible for the practical and social consequences, technical developments, and applications resulting from them. Basic researchers, in particular, saw/see it this way.

In 1994, Nobel Prize winner Rudolf Mößbauer said: "In the field of basic research you have no responsibility at all."[2] However, he added that it would be "different" for applied physics. The Nobel Prize winner Klaus von Klitzing also emphasized that the scientist would only be responsible for the validity of the research results— not for the practical applications by others. And he added: "After all, basic research cannot be banned."[3]

---

[1] Types of responsibility in general and specific responsibility(ies) are analytically understandable (quasi ideal-typical) concepts or, in part, normative constructs of interpretation, which often "overlap" in social reality to the extent that several of the typological constructs are often applied simultaneously—in varying degrees—for description and analysis. This makes clear the interpretive character of the attributions of responsibility in particular. (Nevertheless, the attributions usually have considerable social reality, because they are based on social norms, some of them supraindividually binding or even sanctioned).

[2] Mößbauer (1994), see also Lenk (2015, p. 337).

[3] See Lenk (2015, p. 337).

Albert Einstein, however, was of a different opinion. He wrote to Max von Laue, also a Nobel Prize winner, in the 1930s:

> I do not share your view that scientific man should remain silent in political, i.e., human, affairs in the broadest sense. You can see from the conditions in Germany whereto such self-restriction leads. It means leaving the leadership to the blind and irresponsible ones— without resistance. Isn't there a lack of responsibility behind all that? Where would we be if people like Giordano Bruno, Spinoza, Voltaire, Humboldt had thought and acted like that?[4]

Einstein, as a newly appointed member of the Prussian Academy in the last century, proposed the foundation of a chair for philosophy of physics during a lecture at the "Friedrich-Wilhelm-Universität" in Berlin (now "Humboldt-Universität zu Berlin"). He obviously did not only mean in this later letter just the responsibility for methodologically sound scientific work, but much more generally the purview and "feeling" of a much greater, generally human responsibility—before one's own conscience facing the ethical "moral law" (Kant) and the (idea of) humanity or society—in any case towards an internal and/or external instance.

Accordingly, a clear distinction is rightly made between the so-called external and internal responsibility of scientists. Even today, however, scientists too easily confuse or confound internal and external responsibility. However, moral responsibility—directed towards those potentially affected by actions on the one hand and the traditional *guild ethos* of the scientist on the other hand—should also not be confused here.

The *ethos of* the scientific guild and the "internal" "responsibility of the scientist" are not ethical in the strict sense. Ethos presupposes ethics, but is not ethics. The respective codes of standards of the scientific associations, for example, are in this sense *ethos*, not universal *ethics* of the scientist or even science. Unfortunately, this is often still mixed up. Despite some recent scandals, the ethics of the guild, the ethos system of science, generally works quite well.[5]

The scientist himself usually tends to retreat to the rather narrowly understood *ethos:* Only the best possible, efficient, clean, truthful research ("good practice") and honest, non-deceptive recording and publication, as well as fair treatment of his rivals, etc. would be his responsibility. But this is not enough, for example, when it comes to so-called "human experiments" (experiments on humans, whether performed individually or collectively) or "field experiments" in which people are directly affected or when the transition to applied research becomes fluid. The separation of basic research and applied research has become much more difficult today, sometimes even impossible; Just think of today's genetic engineering.

---

[4] See Herrmann (1977, p. 115, translated). Herneck, a science historian, summarized—also with regard to Einstein's later statements after the atomic bombing of Hiroshima and Nagasaki—as follows: "Albert Einstein is a brilliant example of a scholar who has grasped the problem of the responsibility of the natural scientist and technician in the atomic age in all its depth and is striving to do justice to it" (1977, p. 401, translated).

[5] Cf. Lenk (1991, p. 57 et seq.).

According to Einstein, however, scientists may also bear *external and social* responsibility.

Even the ambivalence of the positive and negative, destructive usability of technical and scientific results can no longer be resolved so smoothly and easily as traditionally thought: also, if responsibility grows with power and knowledge, then the co-responsibility increases accordingly with both.

Is the excuse of the well-known biochemist José M. R. Delgado—namely, "I am not an ethicist, I am a biologist"—generally sufficient to "de-excuse"?[6] A certain co-responsibility of the scientist providing the procedures can be given on a case-by-case basis, which is particularly evident in the negative case: The scientific developer of napalm, Louis Fieser, of course, like the later so-called "father of the hydrogen bomb", Teller, rejected any ethical co-responsibility, although the latter had previously reported his torments of conscience in a letter to Leo Szilard.[7]

Power, ability, and knowledge obligate us. "Everyone has a special responsibility where he has either special power or special knowledge."[8] Karl Popper would like to activate responsibility through an oath-like "promise" oriented on the Hippocratic Oath of the medical profession. It turned out that the idea of the Hippocratic Oath is problematic. It is good as an idea, but has a low effectiveness, (too) low controllability and enforceability. It does not take or enforce enough real political, practical action. It is at best ideal-typical. It may hardly work effectively in field and human experiments. Different rules should probably be used for the application of the results of completed research in social practice.

---

[6] Cf. Lenk (2015, p. 343).

[7] In this letter to Leo Szilard dated July 4, 1945—that is, before the nuclear bombs were dropped on Japanese cities—Teller wrote, "I have no hope of clearing my conscience. The things we are working on are so terrible that no amount of protesting or fiddling with politics will save our souls. This much is true: I have not worked on the project for a very selfish reason and I have gotten mucsh (sic!) more trouble than pleasure out of it. I worked because the problems interested me and I should have felt it a great restraint not to go ahead. I can not claim that I simply worked to do my duty. A sense of duty could keep me out of such work. It could not get me into the present kind of activity against my inclinations. If you should succeed in convincing me that your moral objections are valid, I should quit working. I hardly think that I should start protesting. But I am not really convinced of your objections. I do not feel that there is any chance to outlaw any one weapon. If we have a slim chance of survival, it lies in the possibility to get rid of wars" (Teller, 1945). So Teller only hoped for the deterrent effect. (And perhaps the historical development in retrospect has even proved him right in this respect...). Szilard, on the other hand, relied on the general worldwide publication of the research results and some kind of automatic check-and-balance solution to the problem. Is Teller's statement only impotent cynicism, deportation of all morality and justification (possibly unconscious strategy of self-justification, a so-called rationalization)? The letter rather speaks for conscious moral fatalism or defeatism—as if nothing more could be done. Have scientists and technicians today become the bearers of a pact that is no longer Faustian but downright diabolical, a vicious circle at least, which, as Robert Oppenheimer said, has led them to the edge of the abyss of presumption? Have they now learned to know utmost sin, have they even sinned in doing their research?

[8] Popper (1977), p. 304, translated.

One often refers to so-called "ethics committees." These should not only be used in medicine, but for all sciences. It seems doubtful, however, whether a permanent ethics committee could be in charge of investigating and assessing the ethical, social, legal consequences of basic research and progress not only in biomedical research, but in technology and science in general. How should it be the appropriate institution to steer and reliably scrutinize science, even if this committee were inter-disciplinary and broadly based? It would indeed be absolutely overstressed and overcharged. Ethics committees in biomedical research, as in all direct human experiments, may be useful and in order for monitoring purposes, whereas a com-prehensive overall commission would probably find itself overburdened in dealing with all the overarching problems of basic research. Instead, these issues need to be addressed politically.

However, science must not be unnecessarily hindered or prevented. So far, there are no effective overall remedies for all-round solutions to such conflicts. One should do everything possible to raise awareness of ethical conflicts and not blindly suppress ethical considerations by retreating to career interests, which is indeed a systemic pressure in the unrelenting competition for career progression! (Just think of the German cancer researcher scandal, or the data manipulation by a young German physicist or a Korean stem cell researcher, and most recently Chinese genetic manipulation of embryos).

Incidentally, it is usually not a matter of assigning responsibility solely to indi-viduals, but of (bearing) joint responsibility, of sharing the responsibility in groups, etc. The extended responsibility in view of the Faustian pact on scientific and tech-nological progress, which has been entered into and is no longer easy to revoke, is indeed more important than a traditional moral responsibility for "good scientific" basic research, which can hardly ever be attributed retroactively.

The responsibility of the researcher in science and technology is indeed a special case of a role-specific and moral responsibility in a strategic position. Preventive responsibility must be taken into account wherever harmful effects can be antici-pated and possibly be averted. A personal co-responsibility may exist on a case-by-case basis, but a general strict or even sole responsibility of the scientists and technicians for the causes across all cases does not exist in view of the ambivalence and collective origin of research results, especially in basic research. In most cases it is a question of co-responsibility, which should be specified in more detail. We have to find viable middle solutions. All the more important is the preventive view to prevent destruction and permanent damage in advance, if ever possible. In view of the dynamics of development and the difficulties of orientation and evaluation in this whole problem area, the only realistic approach seems to be to promote the sense of moral co-responsibility as far as possible and to discuss it, for example, by means of case studies.

Responsibility as a Relational Concept of Attribution

A Conceptual and Methodological Overview

Terms of responsibility are ascribable or attributed in the form of multi-place predicates (i.e., relational) or structural terms,[9] schemes requiring analysis and, interpretation with the following elements:

- *somebody*: responsible subject, bearer (person or institution, corporation, etc.) is responsible[10]
- *for*: something (actions, sequences of actions, states, tasks, etc.)
- *towards*: an addressee
- *before*: a (sanction or judging) instance
- *in relation to:* a (prescriptive, normative) criterion
- *within the framework of an* area of responsibility, area of action, etc.

Responsibility is thus initially a concept that is expressed in a relational norm of attribution through the evaluation of a controlled expectation of action. Accountability means that someone has to justify their actions, consequences of actions, conditions, tasks, etc. towards or before an addressee to whom (s)he is accountable and before an authority—according to relevant and extant social, legal, or moral standards, criteria, norms, etc. The person responsible in each case has to justify their actions and decisions, etc., vouch for and be responsible for their own actions and those of others, if specific conditions are met. Moreover, "responsibility" is not only a descriptive term—it is normatively established that someone bears responsibility—to be used, but above all also a concept that can be ascribed in an *evaluative way*: Someone is held responsible, held accountable—which opens up the normative, and thus ultimately the ethical dimension of action. Depending on the type of responsibility, a conventional, social, normative, or descriptive language game is opened or "played." Responsibility by attribution or description is a social construct of interpretation embedded in institutional contexts. Responsibility is relative to the system of norms, is attributed in a context-, culture-, language, system- and theory-related way. (Finally, the functional attribution and disposal methods of the linguistically and socially embedded occurrences of concepts and statements of responsibility would also need to be analyzed more closely).

---

[9] Lang (1985, p. 262), who develops a structural model of legal philosophy, writes that "the formulation of an analytic definition of legal responsibility seems not to be possible" and that responsibility "has many meanings in the different branches of law". However, there is a "core meaning" with some "necessary structural elements": "the bearer of responsibility," "the receiver of responsibility" and "the object of responsibility." See also Neumaier (2008).

[10] Secondary distinctions that do not belong within the concept of relationship itself (as an element) could be: responsible with regard to a point in time: ex ante, ex post; threatened with sanction: formal, informal; with varying degrees of binding force, corresponding to mandatory (must do), target (should do), and optional (can do) norms.

The *attribution of responsibility is thus multidimensional*: it can attempt to determine the causation, the action (consequential) responsibility in a descriptive manner; it can also attribute other types of responsibility in a descriptive manner. Yet, it can also normatively attribute either legal liability and guilt or moral reprehensibility or praiseworthiness. Different types of responsibility are now obtained by (a further) interpretation by a differentiating or specifying allocation of the general scheme of responsibility, the relationship links as mentioned, etc.

To speak of a single concept of responsibility, a single meaning *of the* (total) responsibility, does usually not do justice to the different interpretations, interpretations, or reference perspectives. Different types of responsibility and correspondingly different concepts of responsibility must be analytically distinguished. But they are possibly related, compared, assessed against another, and possibly brought together personally or by "coordination" in "parts" or partial aspects into an integrated overall or "composite responsibility" (see below).

Within the framework of the "usage theory" of meaning (Wittgenstein), Neumaier (1986) examined the concepts of "responsibility" and "conscience". He distinguished different ways of using the concept of responsibility in different meanings depending on which criteria we take as a basis, since we only ever "capture certain aspects" (p. 215). Characteristic of the different ways of use are *family similarities* in the sense of Wittgenstein (p. 217). One can distinguish, among others, the following pairs of meanings (loc. cit.):

- Descriptive and normative use of "responsibility"
- Individual and collective responsibility,
- Also collective and corporate responsibility (cf. Maring, 2001),
- Responsibility for someone who can or cannot assert or uphold certain rights against the actor,
- Moral and legal responsibility.

(The list could of course be extended; see the sections below)

Normative vs. Descriptive Use

With regard to concepts of responsibility, especially with regard to the ability to take over and bear responsibility and characterizing humans as the responsible beings per se in philosophy and empirical social sciences, two aspects should be clearly distinguished: Ethically, the ability to take and bear responsibility is a normative prerequisite, which can be assumed to be virtually independent of experience, in the sense of the moral person's ability to act differently, which is not necessarily meant empirically, i.e., the presupposed freedom for self-determination and for the corresponding imputation.[11] The preconditions are mutually related and interdependent:

---

[11] Depending on the (area of) responsibility, this prerequisite is considered to be given or fulfilled in different ways for (real) persons—if necessary, also graded as in the law (e.g., children who have not reached the age of 7 are not tortable; adolescents who have not reached the age of 18 are only tortable in a limited sense [German law, § 828 BGB]).

the normative definition and its validation leads to the empirical question of the existence of the extant preconditions. Normative specifications are thus based on criteria whose fulfilment can or should be empirically verifiable; in practice, however, it is hardly possible to do something without some form of evaluation.

Thus, within the words of Mackie (1977), it is "factual, psychological, question whether an action is intentional or voluntary" (p. 208). However, it is/will "be a moral or legal question whether or in what ways an agent *is to be* held responsible" (loc. cit., added emphasis). It is also an empirical question to what extent individual actions or certain types of actions meet criteria/conditions of responsibility (cf. p. 215). However, the "straight rule" of attributing responsibility links both problems together: an agent responsible for all and only his/her "intentional actions" (p. 208). We occasionally deviate from this principle: there is—often in exceptional cases or on the fringes of the "family-like" concept areas—also responsibility without intention (e.g., in the case of strict liability) and intentional action without legal and moral responsibility: Thus we consider children "legally and morally less responsible for what they do", even if there is "no general lack of intentionality" in their actions (p. 212 et seq.).

Like Ingarden (1970, p. 5 et seqq., translated), who distinguished between "being responsible", "assuming responsibility," and "to be held responsible" and emphasized the actual "independence of these facts" and their "context of meaning", Ströker (1986, p. 196 et seq., translated) also separated the bearing/having of responsibility and its assumption as well as the context and the normative conditions, "which exist between the individual determinants of the concept of responsibility and their situational moments": Thus, on the one hand, these are "de facto independent of each other": one can have a certain responsibility and yet not take it over and possibly not be called to account. On the other hand, Ströker claims that "one can be held responsible for something without being responsible for it. Also, one can take responsibility without really having it." In spite of this "de facto independence" there are "idiosyncratic" ethical connections: "As soon as one has or bears responsibility for something, one can also, in principle, be called to account and should not eschew one's obligations." Furthermore, "taking responsibility for something for which one does not have, it may well be necessary, but in other cases it may be morally illicit." In this respect, "an abstract general standardization is not possible," but perhaps a more precise definition of the relationships might be.

Types of Responsibility

At least the following types of responsibility can be distinguished, for example (cf. detailed Lenk & Ropohl, 1987, p. 115 et seqq.):

- Responsibility for consequences of *action or causal responsibility* for one action(s); in a slightly modified sense as
- Liability (for damages); then probably rather as a special case of
- *Legal responsibility*,

- *Role or task responsibility,*
- *Moral responsibility,*
- Pedagogical responsibility,
- System responsibility
- (Reflexive) meta-responsibility
- As well as higher-level "*composite responsibility*" with possible overlappings or gradings of special responsibility (types) (see priority rules below).

I would now like to present the corresponding diagrams[12] of the types of responsibility, which I have already dealt with frequently but shall to comment on only very briefly here.

First of all, the fact that someone is responsible for his or her actions or the consequences of their actions can be understood in many ways (see Diagram 2.1). Firstly—and this is the normal situation—it applies that one causes and brings about one's own actions and is therefore (mostly) responsible for them and for the corresponding consequences. This is the (*positive*) responsibility for action. However, there are also omissions, and thus a corresponding *negative* causal responsibility for

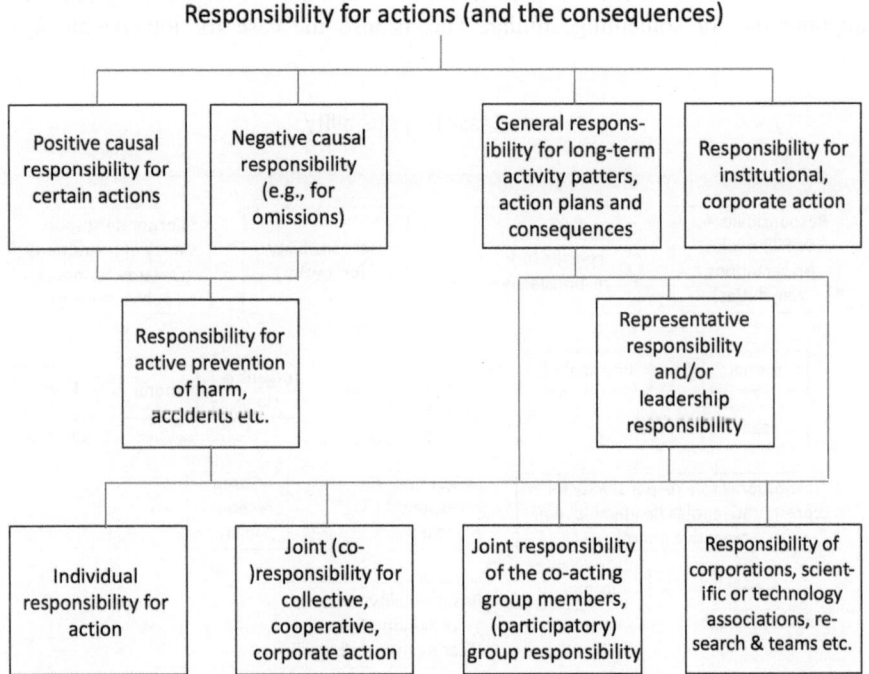

Diagram 2.1 Types of action responsibility (including the responsibility for consequences)

---

[12] First probably published in Lenk (1982), but also, e.g., in Lenk & Ropohl (1987), Lenk (eds.) (1991); Lenk (1992, 1996, 1997a, b, 2006, 2015); and in English in Lenk (2005/2015, 2007, 2019).

action. And there is also the *combination* of both, namely in active responsibility for prevention and protection which, for example, the test engineer or the control scientist must assume in the practice of applied sciences, as does every supervisor in any field whatsoever. This is of course a responsibility that is particularly characteristic of some engineering activities. Then there is also the responsibility for longer-term actions, sequences of actions, series of actions; parents are responsible for their children, for example, etc. Finally, a responsibility for institutional, for corporate action must also be listed, a kind of responsibility that also applies to companies, institutions, or is exercised by representative and leadership responsibility: When one acts as a representative of a corresponding group, society or, for example, a state institution, then one is acting "representatively" in a specific leadership role, as a leader; and this is a kind of responsibility that must be analytically separated from direct personal responsibility. Furthermore, there are of course some overlaps, conflicts, questions of co-responsibility, etc. Diagram 2.1 is of course still a somewhat abstract scheme that needs to be filled in more closely and substantially.

One of the most characteristic substantiations is of course what we call *professional or, more generally, role and task responsibility* (see Diagram 2.2). Everyone who is active in a role has role duties, and must fulfill or execute them responsibly. This can be formal or legal or prescribed; but it can also be informal, by habit, by appointment, or something similar. This is also the case for job-specific task

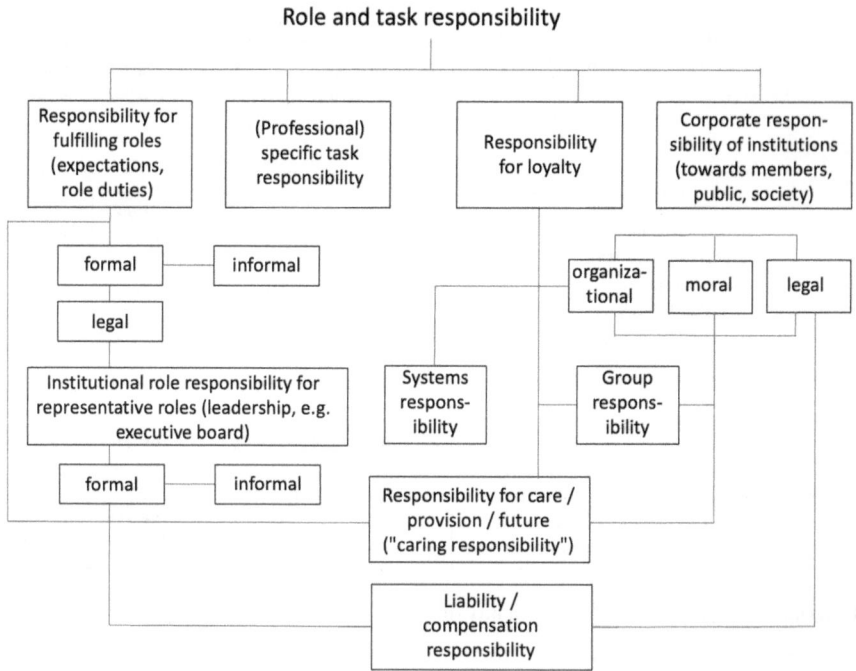

Diagram 2.2 Role and task responsibility

responsibility, for example, which refers to a very narrow job description or similarly defined role. But, independently, there is also a personal responsibility of loyalty, e.g., in politics towards the "elder statesman" and similar personalities or authorities (e.g., the state or the people). (These are responsibilities which are not formally concretized in any way, but that nevertheless exist). There is also, unequivocally, a corporate responsibility of the state towards its members or citizens, and the corporate responsibility of companies, e.g., towards members and customers—both in a legal context and certainly also in the fulfilment of tasks, etc. Yet, there is also the problem of whether such actors can also have a special (*corporate*) *moral* responsibility in this respect.

What is particularly interesting here is that one can also, for example, have a mere liability responsibility, i.e., a responsibility for actions and things that one has not caused oneself, where one is only liable or has to accept responsibility, e.g., parents for their minor children. Worthy of special mention is the responsibility for welfare and precaution, which Hans Jonas focused on in his book on *"Das Prinzip Veratwortung"* (1979, English title: *"The Imperative of Responsibility"*), in which he emphasized an *expansion of* responsibility (or the concept, respectively). Jonas considers that the traditional concept of guilt responsibility should be abandoned and instead be replaced by an expanded concept of responsibility in terms of power or existence dependency: Children are dependent on their parents and the parents are responsible for the dependent children in *general*. And so in general the more powerful person is always responsible for the dependent one. Of course, it is not justifiable that this is now *the* responsibility that is supposed to replace the "old" guilt responsibility, as Jonas had originally claimed. He then realized in subsequent discussion with me that this must be changed: The traditional responsibility for one's own actions, i.e., for one's present and past well- and wrongdoings etc. as well as for one's own future actions (Jonas: "that which is to be done") naturally remains. But according to Jonas, the *responsibility for care* is indeed an ethical extension (or at least necessary accentuation) of responsibility. Incidentally, at the same time as Jonas emphasized that shift, I had already emphasized that "extended possibilities for action" also generate "extended responsibilities" (Lenk, 1979, p. 73).

Of course, one could give many more examples, especially from science and technology; I do not wish to do that here, but only refer to the examples mentioned at the beginning.

Pure *moral responsibility*—I refer instead to *universal moral responsibility* in order to distinguish the real ethical responsibility from the extant moral(e) (which, for example, the Mafia also has in its codes of conduct, as is well known, and a very strict one)—is that which applies *equally* to *all* in all comparable positions and situations (see Diagram 2.3). This is often activated by direct situations, action situations, decision situations, etc. Here, also the previously mentioned responsibility for care and precaution for one's dependents in the sense meant by Jonas naturally also comes up again.

However, there is also *indirect* responsibility for the possible consequences of one's actions or omissions, for example, as remote consequences. For example, there are relations between highly industrialized countries; for example, in the case

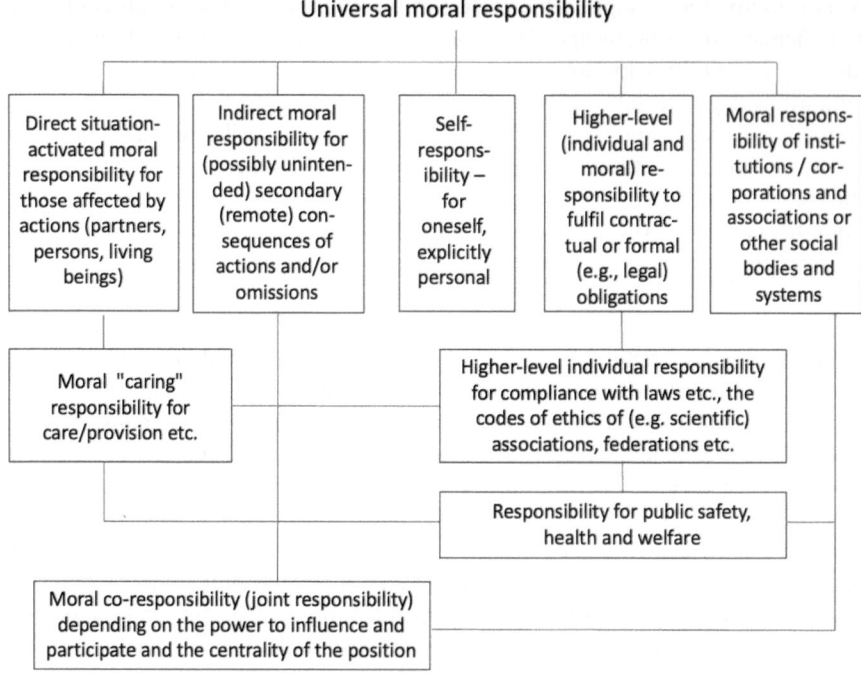

Diagram 2.3 Universal moral responsibility

of coffee prices and coffee producers in developing countries, there are economic survival problems that lead to livelihood problems for coffee farmers, etc. These responsibilities here are quite similar to the corresponding role and task responsibilities, only here they are related to the truly ethical problem, to that which concerns the welfare and woe of other persons—or even other living beings, e.g., pets—and thus concerns the ethical–moral proper.

There is also a *higher level of* individual responsibility to fulfil certain moral or other contractual, legal, or formal obligations. For example, I have a moral obligation to abide by the law. This is a higher-level moral obligation, such as the obligation to comply with special laws, to assume certain subordinate or lower-level responsibilities. The question of whether there is a moral responsibility of institutions, companies, or corporations is a hotly debated one. I mean that such a responsibility indeed exists. But I hold that this responsibility cannot be understood in the way that some American scholars (e.g., French, 1984) present it; they see the corporation as a "moral person" comparable to a legal person. But this is something that can and should be talked about and discussed.

An important passage for technicians and scientists is the following: There is a responsibility to comply with the codes of ethics and the corresponding standards that make up the ethos of the respective associations and those that relate to responsibility for the general public: in the US since 1947, all codes of ethics have included the responsibility for maintaining or ensuring "public safety, health and welfare."

In law, the situation is quite difficult because different areas of law have different concepts, including tort and liability (and not only, for example, family relationships in the case of paternity or inheritance, which vary between different areas). We have not yet found a really clear picture of the typology of legal responsibilities. However, we can say: A norm addressee is responsible to the authorities to which (s)he must answer, depending on subjective preconditions, graded according to degrees of responsibility and with various legal consequences, in particular sanctions. For example, the following responsibility-relevant elements can be found in German law:

1. Addressee of the norm: e.g., in criminal law: individuals; in civil law: natural persons and legal entities for their organs.
2. Instances: e.g., in criminal law: courts; in civil law: individuals, arbitration bodies, courts.
3. Subjective requirements: e.g., culpability, intent, negligence, warranty for purchase and work contracts, (strict) liability.
4. Characteristics of responsibility: e.g., behavioral responsibility for action/inaction (or intentional omission); role responsibility: e.g., contractual, parental custody, restricted or specifically limited responsibility.
5. Legal consequences, in particular sanctions: e.g., without sanctions (fully / partially released from responsibility); with sanctions: positive (tax benefits, subsidies); negative (liability, penalties).

In most cases, even in the case of engineers and scientists—especially in applied research—responsibility refers to specific roles and often to *conflicts* between such roles, duties, and expectations and their various responsibilities, and to corresponding distributional issues. The engineer as a person has to deal with different corresponding institutions, e.g., with clients, customers, or employers; one's own company or a foreign one, etc.; with corresponding public institutions or the profession itself, i.e., the respective association etc.; or with society in general.

Accordingly, *conflicts* can naturally develop between different loyalties and responsibilities; this is even quite typical. Such a conflict of responsibilities arises, for example in situations that remained common during the Last century, where a company or employer might demand of an employed engineer that this subordinate coworker should dump waste into the Rhine river or into the air, for example at night, which is of course contrary to the interests of the public and (more recently) of course illegal under environmental law. Such a situation naturally leads to a personal conflict of responsibility. What is the poor engineer supposed to do? The conflict may be difficult to bear or to solve—sometimes with serious consequences for the employee. That is why we have considered whether there are certain regulations or possibilities to address such conflicts. I would like to try to shed light on this with the following rules for listing priorities and preferences.

Certain ideas for conflict resolution may be adopted from American business ethics (these are the first four rules): In essence, these first rules say that there are basic moral rights, in particular Human Rights, that are inviolable; even our Constitution already says this. In addition, there is a plausible demand that considerations of benefit and acceptable livelihood and public as well as private health and societal Commons must be taken into consideration in referring to these fundamental rights—in particular if there are insoluble conflicts between fundamental rights or some extant equivalent rights. For example, fair compromises should be sought: after weighing up the moral rights of each party, some kind of compensation should be found, with some kind of decent proportionality. Only after these rules have been applied should one weigh the anticipated benefits against those of doing or causing harm to others. This is a consideration that frequently occurs in business ethics and can be summarized in such a way that one should take into account non-surrenderable moral rights before averting and preventing harm and these before considerations of benefit. In practically insoluble conflicts, therefore, an attempt should be made to achieve an equal distribution or "fair" proportion of the corresponding distribution of burdens and benefits. Universal moral responsibility should therefore generally precede task and rule responsibility or role responsibility. The public good should take precedence over individual interests.

Box 2.1: Twenty Priority Rules[13]

1. Weighing up the moral rights of each individual concerned; these take precedence over considerations of benefit (predistributive, basic rights).
2. Seek a compromise that takes everyone equally into account; in the case of an insoluble conflict between equivalent fundamental rights.
3. Only after weighing up the moral rights of each party can and should one vote for the solution that causes the least damage to all parties.
4. Only after 'application' of rules 1, 2 and 3, then weigh up benefits against doing or causing harm. In other words, *moral rights that cannot be abandoned take precedence over averting and preventing damage and the latter over considerations of benefit.*
5. *In the case of practically insoluble conflicts between parties and those involved, certain fair and humane compromises should be sought with regard to the harm and benefit to the various parties.* (Fair compromises are, for example, approximately equally distributed or justifiably proportioned, sharing burden and benefit).
6. Universal moral and direct moral responsibility takes precedence over non-moral and limited obligations.
7. Universal moral responsibility usually takes precedence over task or role responsibility.

(continued)

---

[13] The first four rules are taken from Werhane (1985) , pp. 72–3.

Box 2.1: (continued)

8. Direct primary moral responsibility in the action or decision situation is usually given priority over indirect remote responsibilities (because of the urgency and limited obligation; but: evaluations and gradation are necessary according to the severity and sustainability of consequences).
9. Universal moral and direct moral responsibility take precedence over secondary, e.g., corporate responsibility.
10. The public good—the common good—is to prevail over all other specific and particular non-moral interests.
11. In the case of safety-related design, preference shall be given to the solution by which the protection goal is best achieved in a technically sensible and economically viable manner. In case of doubt, safety requirements take precedence over pure economic considerations.[14] Safety therefore comes before economic efficiency.
12. *Global, continental, regional, and local environmental compatibility must be distinguished and taken into account. System-relevant / decisive environmental compatibility takes precedence—and, in this extreme type, the more sectoral or comprehensive* (cf. the climate crisis).
13. *Ecological compatibility and sustainability take precedence over economic use,* except in cases of immediate "urgency" (e.g., famine, epidemics or even pandemics, or other humanitarian disasters).
14. *Human, humane, and social compatibility take precedence over environmental, species- and nature compatibility in individual cases of conflict,* but are usually to be *striven for* together or *in sensible compromises.*
15. Concrete humanity takes precedence over abstract demands and universal principles (concrete humane and socially acceptable weighing of goods).
16. Humane (human and social compatibility) concerns take precedence over the merely factual.
17. *Compatibility with the requirements of survival and the quality of life of future human generations and the predictable acceptance of measures affecting future generations should be given very high priority.*
18. In social and political planning in general, every effort should be made to achieve a (relative) maximum of general freedom and freedom of choice—openness and flexibility of planning on a large scale—and to achieve largely equal opportunities for future developments ("multi-option society").
19. *A relatively wide range of options should be given high priority for present and future generations, i.e.,* no important options should be excluded for present and future generations. *It is therefore necessary to avoid total*

(continued)

---

[14] This rule refers to technical regulations by DIN 31 000 (ISO 51: 1999).

Box 2.1: (continued)

resource depletion and extensive environmental pollution by giving priority to "sustainable development" everywhere, which neither overtaxes nor undermines the "load-carrying capacity" of ecosystems (and especially of the entire ecosystem of the sphere of life)—nor runs counter to the fundamental rights and participation rights of large population groups. It is therefore imperative to achieve a proportionate and morally acceptable combination of the requirements or priorities of Rules 16 to 18. Rule 19 refers to the compatibility of today's so-called "multi-option society" and the "sustainable" use of natural reserves and resources without overexploitation—both for present and future generations. In a way, it combines the two rules mentioned above into a demand for a balanced and fair compromise for all parties concerned—both present and future. *"Sustainable development" for present and future generations should be a very high priority. The idea of the "sustainable" use of resources with renewable raw materials of all kinds, that has met with great international acclaim (even if it has not yet been realized), should therefore be further supplemented by the demand for the non-total exhaustion of non-renewable resources and by the search for alternatives that are as environmentally friendly as possible. In particular, highly important organic raw materials such as petroleum, for example, which should still be available for future petrochemical syntheses and developments and thus for important products for future generations, should not continue to be burned uncontrollably by combustion engines in ever in-creasing amounts.*

The current possible freedoms of multi-option societies must be preserved in an appropriate manner and, if possible, also for future generations, and that they must be given access to a reasonably distributive assurance of the conditions of existence (the minimum standard of living beyond the physical subsistence level). Not only a "natural" but also a morally "acceptable," i.e., humane and humanitarian, form of development should urgently be aimed for. A combination of the latter two rules should be sought for both present and future people in an appropriate manner as emphatically as possible and as "sustainably" as possible.

In cases of urgency, ecological compatibility and/or sustainability should usually take precedence over purely economic interests and application.

Finally: *Concrete humanity*—the concrete-human combined responsibility mentioned above—takes precedence over abstract demands and universal principles. This means that the corresponding decision in an urgent or emergency situation comes first: here, I would say, the human or humanitarian benefit or responsibility for the people involved even takes precedence over environmental compatibility in the broader sense, although the two are generally closely related in the end.

The Necessary Personal Integrated Balance of Responsibilities

Concrete-human responsibility, including for the consequences of developments in complex systems, can neither ethically nor legally be borne out by a single individual alone. Of course, this also applies to the ethics of technicians and scientists. However, it cannot be assigned in an abstract way to the human *species as* such or to the professional category of engineer or manager. Yet, medium solutions according to the respective situation or role are, as a rule, dependent on the centrality, decision-making power, or potential impact, and on group responsibilities, etc. They are to be developed in a graduated sequence or, in the case of conflicts, dimensioned according to the viewpoint and imperative of concrete humanity. No one can be responsible for everything. Responsibility is not all-encompassing, especially not in the age of networked systemic contexts, where the problems of distribution of responsibility and multiple allocation are particularly difficult; For example, who is responsible for the information on the Internet?[15] But neither can we ignore humanity, humanization, in the form of really humane measure(s). That, though, is all too easily forgotten in these contexts. And humanity becomes effective only in concrete terms. *Humanitas concreta praestet!* (Concrete humanity has priority).

In summary, I would like to assert that personal moral responsibility continues to be the prototypical example, the model of responsibility. But that personal responsibility is no longer the sole responsibility. Although individual moral responsibility is the prototype, there are also responsibilities of collective actors and of formally organized secondary actors (institutions, corporations, enterprises etc.), i.e., a secondary responsibility, so to speak, for organizational, corporate action, which must, however, *always* be seen *in connection* with personal responsibilities, with the "ethics of the personality" to which it cannot, however, be reduced completely. It is and remains a difficult, precarious problem to keep alive the connection between the more abstract organizational levels from the "ethics of society" with its implied and possible systemic or structurally engendered "inhumanities" on the one hand and the concrete, personal responsibility in real situations and in the case of overlaps and conflicts on the other hand. Both are considered to be particularly important in terms of concrete humanity. So, we always have to perform a difficult balancing act, especially when dealing with institutional and corporate responsibilities.

Here Are Some Summarizing Theses

20. It is important that emphasizing collective or corporate responsibility should not serve as a "shield" or maneuvering trick to *distract from individual personal* responsibility, thereby opening the door to personal irresponsible action—in the sense that we would, for example, claim that individuals are no longer personally responsible, but only the state, institution, group, or society at large. We

---

[15] Is anyone here to be held responsible in a tangible and controllable way? Developing an information ethics in this respect is a very urgent task for the near future; in fact, one does not see any possibility at all for the concrete shaping of an operationalizable ethics with regard to the worldwide information systems—except for the necessity of expanding the traditional concepts.

know the excuses of the "cogwheels," the excuses of the concentration camp henchmen for following emergency orders and the like.

21. It should be noted and emphasized that supra-individual responsibility does not become obsolete simply because certain individuals bear some level of (co-) responsibility. There are indeed collective and corporate responsibilities that cannot be reduced to the individual, although they are always connected to the personal in the sense: Even in view of this irreducibility, some form of personal co-responsibility is always activated in a specific case.

22. Wherever possible, responsibility should also always be understood as open for participation and open to the future (i.e., for the control of *future* actions, decisions, plans, risk apportionment) and cannot merely be reduced to assigning blame for past actions to individual scapegoats: Responsibility in systemic contexts and action and decision-making structures is always an essential part of the practice of joint or co-responsibility and responsibility for the future. Individuals cannot be held *solely responsible* for what they have not caused on their own, nor for events for which they cannot be fully responsible. But as participants or members, they can bear co-responsibility—to the extent of their participation, power of influence, or participation in decision-making or the centrality of their position.

23. According to Jonas (1979, Engl. 1984), this responsibility for the future always includes not only the ethics of precaution and prevention, but also the *responsibility* for care—especially for those who are dependent on us, in accordance with the situation and social situation.

24. It is also particularly important to note that the attribution of individual, personal responsibility must always be seen or embedded from the perspective of concrete humanity and its dimensions. Albert Schweitzer's "ethics as concrete humanity" (cf. Lenk, 2000) is and remains a prototypical model for this.[16]

25. Thesis 5 is important, especially in view of the increasing predominance of associational and institutional powers and influences that threaten to displace the individual and his or her contributions and influences. This also applies within institutions, in technology and economy: in case of doubt, defer to concrete humanity!

26. Only concrete humanity can make the general idea of humanity tangible, make it operational, keep it bearable in the sense that the extant concrete responsibility is simultaneously appropriate to the situation, open to participation, and prospective.

In dubio pro humanitate concreta!
In case of doubt: defer to concrete humanity!

---

[16] Although Schweitzer wrongly devalues the collective responsibilities of "social ethics" and of organizations, institutions, and groups etc. as not actually ethical.

Fig. 2.1 "We should divide the burden more fairly." After the German version in Badische Neueste Nachrichten, Karlsruhe, (repr. in Lenk, 2019, p. 279, reproduced with permission)

Distribution of Responsibility(s)

How should responsibilities be divided in a fair and plausible way? I copied the following picture once from the *Badische Neueste Nachrichten* (Fig. 2.1):

The distribution of responsibility in this form may not be the (patent) solution to all problems, but a distribution problem will of course always arise—or often even several distribution problems. These are often quite difficult to treat. I would like to mention just a few theses on this subject:

27. It cannot be assumed that collective responsibility can always be reduced/ reduced completely to the individual, personal responsibility of the actors or defined solely by them. This is ultimately easy(ier) to do, namely to divide or measure quantitatively, only in the case of compensation obligations.

28. It is necessary to develop an extension of responsibilities through operationally manageable, functioning models for the distribution of *co-responsibility* or co-responsibility.

By no means are mere appeals enough to avoid conflict situations or to avoid possible "social traps" that can arise (e.g., from the structure of the so-called "prisoner's dilemma"). There are contradictions that are, so to speak, embedded in the situation; for example, ecology (overgrazing of pastures in the Sahel already treated by Hardin in 1968) can be used to illustrate this very vividly. This means that you need for practical application more than analyses and appeals—yet indeed both are necessary. You have to develop operational access options—and this is at times cumbersome; but it is also difficult to carry out them in adequate detail. As a guideline, for example, one could often use a rule familiar from economics: Only as many laws, commandments, orders, and prohibitions as necessary—but as many incentives, personal initiatives, and personal responsibilities as possible.

A dilemma of responsibility exists, and often arises when committees make decisions, in so far as the anonymity or protection of committees can dilute certain decisions and the responsibility of the individual seems to disappear, so to speak. This is still the case today, even in Parliament, which is why there are sometimes personal or roll-call votes.[17]

On the Question of Responsibility, Especially in the Applied Natural Sciences

First of all, to get you in the mood, here is an example of the discussion of responsibility among some scientists.

It was 1984, i.e., still in the *pre*-Chernobyl era, when I wrote in our university magazine "Fridericiana" in an essay on "Responsibility and Technology" that the responsibility for major scientific and technological projects could no longer really be borne by individual persons: "An individual could only bear the responsibility for a major technological project pro forma, in form, publicly—politically, as it were. But what good is it if he (e.g., the manager of a nuclear power plant) resigns after an MCA (maximum credible accident)—after a major accident to be assumed? Mere formalistic assumption of responsibility no longer seems sufficient." I then received an angry letter from a physicist based at the Jülich nuclear research facility, from which I would like to quote: "What Professor Lenk says about the real aspects of responsibility, especially that of the technician, for example at the point where there is talk of an MCA (maximum credible accident) is, to put it mildly, the worst distortion of the facts." The MCA had only (?) as a "design-basis accident" binding reality for the technician. The responsibility of the technician for ensuring that such an

---

[17] Interestingly enough, this had even been observed in American expedition groups climbing Mount Everest, which were subjected accompanying to social-psychological research. It was found that the decisions that the groups each made were riskier than the decisions that individuals would make if they were solely responsible for them, although group decisions sometimes involved risking the life and death of the respective members. This is an interesting phenomenon called the "risky-shift" phenomenon (according to Stoner, 1968). This is found in many such or similar situation structures, quite apart from the usual problem of "dilution" of responsibilities found in committees.

incident is reliably controlled "cannot be eliminated by ignorance; it is codified in law and can be prosecuted." He also noted: "but it is about the factual sides of responsibility of the powerful of the word, e.g., for what they want to do, operate, suggest, or set in motion with the mere word MCA." The author of the letter also complains about the traditional academic treatment of "the ones of Professor Lenk's caliber" dealing with traditional ethics and says that "there are still considerable deficiencies to be remedied." He advises me "to deal less with the ethical problems of technology and more with the various techniques of responsibility, to deal with them effectively in journalistic terms, for example under topics like these: 'Responsibility and ignorance', 'Responsibility and modern politics', 'Responsibility of academic teachers of today'." The physicist is right with the latter advice. And I was glad to take it—indeed, I did so in rather many publications and practice-oriented seminars together with several colleagues from departments of technology. However, he did not understand the point of the argumentation at all. Perhaps he should not be reproached with the fact that he regarded the MCA, the maximum credible accident, merely as "design-basis accident," i.e., as not realistic, but only as a model fiction (it was not yet publicly known that a core meltdown had also previously occurred in Harrisburg)—and certainly not as realistic or realizable regarding the "super-accident," which goes beyond the *assumed* model *to be assumed.* Perhaps more importantly, political strategies of individual responsibility (the chairman of the Chernobyl nuclear power plant *was,* as is well known, dismissed) and legal codification are in fact no longer sufficient as instruments of regulation. The scapegoat search and perception is more like a ritual of displaying powerlessness. One (and *only* one) should be responsible, held accountable. Science and technology have apparently become too powerful to be adequately covered and even, in extreme cases, controlled by the traditional measures of political and legal regulations of purely personal responsibility—especially in the sense of reprehensibility (Bodenheimer, 1980; Ladd, 1990).[18] If the chairman of a nuclear power plant or the responsible minister has to resign (or—more realistically in Germany—sends his state secretary into early retirement with a not inconsiderable pension), then indeed this only actually shows the relative powerlessness of such regulations. I repeat: the problem of responsibility can no longer be solved in a merely political and formalistic way in view of the major projects of the great power science and technology and its societal impact and factual power of influence.

To whom are scientists responsible? To their individual conscience only? But isn't conscience rather a medium, a "voice" of self-attribution, of self-responsibility, in other words, an instance that estimates and measures responsibility, that applies a criterion that already presupposes itself? Is moral–practical reason this decisive instance, as it has always been seen in the tradition of philosophy, especially with Kant? Or the idea of human self-esteem, the "idea of humankind" or of society? Are we ethically responsible to humanity or society or the law? Well, all of that in a

---

[18] Rather, the (negative) formulation of the strategic responsibility for prevention and precaution according to H. Jonas (1979) seems to be fruitfully open to sharing responsibility and co-responsibility, without the overall responsibility or that of the individual participants dissolving.

sense, yes. But these are all abstract concepts, not living personal entities, not partners whom someone could hold directly responsible and accountable. Responsibility
towards an abstract or an idea remains a metaphor, however effective it may or
should be. Social controls or legal controls concretize this general social responsibility, but they are already decreasing in relation to direct personal ethical responsibility. In particular, it must be said that individual ethical responsibility is ultimately
always directed at a *person*. Final responsibility is *personal* responsibility[19]—thus
says the tradition. But, as we saw, that is not the only responsibility we have to deal
with: Individual responsibility is important, but not enough. Ethical responsibility is
more than the empirical voice of conscience. Societal bearers of responsibility are
also very important and to be taken into consideration—and their social responsibility is to be somehow operationalized into the practice of technological control, planning, and leadership as well.

There is no doubt that both the ideas of social and personal responsibility are
closely related to the idea of human dignity—the dignity of both the human society
at large and that of persons responsible and the (human) addressees. It is part of
human dignity, of the corresponding obligation and of being human, to assume and
exercise responsibility, provided one is an acting and relatively effective free being.
Freedom of action and responsibility are mutually dependent. The idea of human
dignity encompasses that of respect for one's fellow human beings and for one's
own person as well as for human groups and institutions, etc. That would certainly
include the idea of existence and human(e) survival and the development of humanity, which was particularly emphasized by Jonas (1979/1984).

Furthermore, I think it is part of the idea of human dignity that we—as insightful
beings who can at least partially recognize, decipher, and partially direct and control
the natural (eco-)systems and phenomena—should preserve and take care of them
respecting their own sort of "dignity." We can and should also take responsibility for
other natural creatures and even for comprehensive natural systems (ecosystems).
This responsibility grows with our insight and our ability to intervene, especially
with our enormous destructive power—which in that regard is in danger of "running
wild" in several ecological respects. We can and should, as insightful beings, think
representatively for other beings, know ourselves responsible and co-responsible
for them as well if they are dependent on us.

One may then ask how, in view of the diversity of the concepts of responsibility
mentioned, it is possible to arrive at concrete, globally uniform decisions that are
nevertheless appropriate in terms of problems and situations, in a humane or care-
taking way. After all, our traditional intuition is that, ultimately, responsibility must
somehow be indivisible, at least—but not only—as far as personal responsibility is
concerned. Even community responsibility, in a certain sense, is, I would not say
divisible in the sense of distributable or divisible, but rather *open to participation*,
such as the responsibility of members of parliament. Such overriding responsibilities affect everyone almost alike and cannot be minimized by division. All that must
not and should not—especially in the moral sphere, but also in parliament, for

---

[19] Again, we find ourselves being led back to Kant's approach of the ethics of the *Categorical
Imperative* ("*Act representatively!* ", i.e., "act in such a way that all should want to act in such
a way").

example—be based on the socio-psychological principle that the more people work in a group and actually bear responsibility, the less the individual feels that they bear personal responsibility. So, unfortunately, there is a dilution effect, which has already been examined philosophically and analytically, without all these problems having been solved.

Natural scientists, especially physicists, usually have an easier time than social and human scientists because they do not experiment *directly* with people. However, they also often make it easier for themselves, sometimes too easy. Nobel physicist Rudolf Mößbauer answered the question of what he thought about the responsibility of the natural sciences:

> This is very much in evidence, especially in Germany. In the field of basic research, one has no responsibility at all. We try to understand how nature works. It is something different when you do applied physics. But even that is exaggerated excessively in this country. I'm thinking about reactor technology… You simply cannot ban science. And if we stop science here in Germany, it will continue somewhere else. In Germany, hostility towards science is steering the entire research landscape into a very critical situation.[20]

Von Klitzing, also a Nobel physicist, also said that "in the application of research results" the natural scientist would have a responsibility: "In basic research this is not the case, after all one cannot forbid research."[21] The question of external responsibility in basic research is, of course, a serious problem, which has a tradition especially in physics—and not only in applied physics. This has been well known since the Manhattan project, namely that of developing atomic bombs, and the problem has been much discussed. But science had lost "its innocence" much earlier (Herrmann, 1982). One would at least have to refer to the development of combat gases by Fritz Haber who, as is well known, planned and promoted the first German use of poisonous gases in World War I (contrary to existing co-responsibilities under the Hague Convention, to which Germany was already a signatory) and continued some relevant research even after the war (!), together with other well-known scientists: Otto Hahn was also in this group, as were Richard Willstätter, Hans Geiger, and also James Franck who later even drafted and presented the Bethe Franck Report *against* the use of the American atomic bomb on civilians.

Through all these projects and experiences, the *external* moral responsibility of scientists has of course become the subject of discussion in varying degrees of detail.[22] As already mentioned, even among scientists it is easy to confuse the *internal* and *external* question or form of responsibility. Ethics of science, or more precisely: general ethics or universal morals in the sciences or moral responsibility towards the potentially affected persons on the one hand and the guild *ethos*[23] of the

---

[20] See Lenk (2015, p. 337)

[21] loc. cit.

[22] It should not be denied that the majority of research in physics and chemistry at least does not show the extreme escalation of external responsibility, as discussed here using exceptional examples. The following—admittedly extreme—examples are discussed in order to raise the profile of the problem of external responsibility, which is rarely encountered in everyday research and may even be suggested. (cf. the examples of Fritz Haber's initiative in the gas attacks of the First World War and the mentioned topic of the development of atomic bombs).

[23] The "rules of conduct" for scientists, which Mohr (1979) established, are such *ethos rules*.

scientist on the other hand should not be confused, although they should/must always be combined in the practical conduct of science. But therein lies the very problem.

I do not want to discuss the internal responsibility in the ethos of science in more detail here, but I will examine in some detail the external problems of the responsibility of the scientist.

So scientists also bear external responsibility. We want to discuss this in relation to those possibly affected by the results, for example directly by the research process. As already indicated, scientists tend to retreat to the *ethos* and say that only the best possible, efficient, clean, truthful research and honest, fair treatment of rivals would be their only "real" responsibility. But of course, this can no longer be true when it comes to direct human experiments or field experiments in which people are directly affected. Nor can it be taken too lightly if and because the transition from basic research to applied research becomes fluid. Just think of today's genetic biology, where the two can no longer be separated in a real and meaningful way. At minimum, the separation in detail has become very difficult. Of course, Lübbe (1980) is right in a certain sense when he believes that the scientist is overburdened with the full responsibility and the imposition of the assessment of all "harmful side effects" of scientific and technical progress. "Only bottomless moralism," he says, whose "responsibility pathetic is only the complement of its practical impotence," can extend the responsibility of persons beyond their power to act. But in reality, this is no longer possible in the barely penetrable forest of possibly grossly ramified responsibilities. Humans have simply become too powerful with their instruments, their scientific technology, their major interventions in eco-systems[24] in order not to feel jointly responsible for the impacts on humans and for the overall context. In view of the existing dangers, however, it is not enough to take overstraining insights as a reason to sit back and relax. This applies in principle and on a case-by-case basis also to individual scientists at strategic points in the development, application, and implementation of experimental research projects. I believe that a much more differentiated approach is needed here, and that the political and ethical problems must be integrated—including legally!—to tackle them.

Even the ambivalence of the positive and negative, destructive usability of technical and applied-scientific results can no longer be resolved so smoothly and easily by a Gordian knot sword stroke, in that or rather by the simply irresponsible basic research and the applied research for which common responsibility is to be taken. Often, neither could be completely separated from each other. All this has become much more difficult today. Not *feeling* responsible easily turns into irresponsibility. And if—as is our intuition—responsibility grows with power and knowledge, then the co-responsibility *of the* human being in general and of the powerful and knowing individual increases accordingly with both.

---

[24] Most of which have already become "artificial," "technogenic" small worlds with but residual nature.

So what is the external responsibility of the scientists or the researcher? The problem was made particularly clear again by Einstein, especially in his letter (actually written by Szilard) to President Roosevelt, in which, on the advice of Szilard and Wigner, Einstein recommended with a "heavy heart" the development of the American atomic bomb. Or later, after the bombs were dropped on Hiroshima and Nagasaki by the Atomic Scientists of Chicago and by the Society for Social Responsibility in Science, which was founded in 1949 together with Einstein and Paschkis.

The German branch of the Society for Social Responsibility in Science was founded only much later, in 1965. For the Federal Republic of Germany, however, one should also think of the call to action by the "Göttingen Eighteen" group of nuclear physicists, which ""was borne by the recognized practical co-responsibility of the knowledgeable person in a strategic position. Indeed, it was politically effective against the potential nuclear armament of the German Armed Forces. Or we should think of the first Pugwash Conference on Science and World Affairs in 1957, whose co-founder Rotblat was awarded the Nobel Peace Prize in 1995. This primarily morally motivated commitment of the scientists was later institutionalized in the Association of German Scientists, but did not lead to a broad general ethical debate but rather to concrete criticism and project evaluations, sometimes with some political explosiveness.

The Nobel Prize winner Max Born, one of the Göttingen Eighteen,[25] was extremely pessimistic:

> In our technical age, science has social, political and economic functions. No matter how far removed from technical application, one's own work is a link in the chain of actions and decisions that determine the fate of the human race. This aspect of science came to my attention in its full impact only after Hiroshima. But then it took on overwhelming significance and made me think about the changes that the natural sciences have caused in the affairs of people in my own time and where they might lead. Despite my love for scientific work, the result of my reflection was discouraging. It seems to me that nature's attempt to produce a thinking being on this Earth has failed. The reason for this is not only the considerable and even growing probability that a war with nuclear weapons could break out and destroy all life on Earth. Even if the catastrophe can be avoided, I dare to see only a gloomy future for humanity (1965, printed 1969, translated).

Born believes that the real disease of our technical age is the "collapse of all ethical principles." All attempts to adapt our ethical code to our situation in the technical age have failed. In my opinion, however, we cannot speak of such a "collapse" of *all* ethical principles; Rather, relative ineffectiveness, especially in the international arena and with regard to the technical possibilities of impact. Why is that so? How can new ethical orientations be gained that are appropriate for our systems-technological world of today? One thing is clear: we cannot afford, now or in the future, to neglect the urgent ethical problems of science, especially applied science and also technology.

---

[25] In April 1957, a group of 18 scientists published the Göttingen Declaration, which expressly opposed plans for nuclear armament in the Federal Republic of Germany. (see also Mieg, in this volume, Chap. 3).

In experiments with humans, so-called human experiments, people are directly involved in the scientific process during the research process, become, so to speak, objects of research: human guinea pigs—sometimes against their will and knowledge. The external responsibility is also particularly clear in so-called "field research" and applied research. However, it is not limited to this area. Opinions on this external responsibility are still very far apart. It has been said, for example, by the biochemist Ernest Chain (1970), that science as a descriptive study of the laws of nature has no ethical or moral quality, but is instead ethically neutral, and therefore, according to Chain, the scientist cannot be responsible for *any* harmful effects of his inventions—but, if anyone, then *society* is responsible. To society, of course, every scientist is obliged as *a citizen*. In particular, the scientist would not be—according to Chain—responsible for the application by others of a fundamental law which they had discovered and of whose applicability they could not even suspect at the beginning of their project. To hold them responsible for their discovery is tantamount to demanding that they correctly anticipate the outcome of their investigation before it has begun. The decision to pursue a particular application of scientific knowledge goes—and this is correct—far beyond the descriptive knowledge. It would therefore be pointless to ascribe to the scientist a responsibility for the application of their discovery in ways that were not decided by them self. The politician or decision-maker *alone* would have to take this on as *their* responsibility. Chain even goes so far as to say that scientists and technicians engaged in military research into the development of new weapons, whether ballistic or biological, have no responsibility for the terrible destructive effects of the weapons they develop. On the other hand, it has been emphasized, e.g., by Belsey (1978/1979), that, although at first glance freedom of research seems to be a general principle, there are nevertheless restrictions and special responsibilities in view of dangerous research areas, which include, for example, special risks for humanity: especially if the scientist them self has good reasons to believe that their discovery can be used by a political decision-making body in a way that is harmful to humanity, and that, for example, a government would probably use this development in such an abusive manner. In this case the scientist should not put this discovery in the hands of the government. Then, the scientist cannot (and this will probably be particularly explosive in the field of biotechnology and genetic engineering) simply wash their hands clean in public when discovering something that could be disastrous for mankind. Of course, one cannot demand that the scientist be able to correctly predict the outcome of research before it begins, but one can demand that (s)he would and can estimate probable disastrous results in some high-risk areas of research and should evaluate them in the overall framework and make a balanced assessment. But this is part of his/her normal human responsibility: Belsey says that there is no need for special morals in the ethics of science, but yet the scientists and technicians who apply them are occasionally at strategic junctures of the decision, which bring extra-technical and overarching connections into play and demand that the possible consequences of the decision be considered, even if is it only possible to obtain a partial overview of these consequences in advance.

The responsibility of the scientist is limited to the support of interdisciplinary cooperation and the timely and comprehensible information about scientific discoveries and about new technical possibilities and their problems. That also refers to the participation in pilot test projects. To wit, a university colleague postulated—without much irony—an appeal to the humanities to at last now take as their duty to "make *their* moon landing" ("we have done ours!") and to "solve"(!) convincingly the ethical problems of the applied sciences. A "solution" then to be adopted by the natural scientists and technologists. An all too easy piece of advice to shuffle off (thinking and realizing) one's own responsibility for one's professional activities.

Karl R. Popper said that "*only natural scientists*" could, for example, foresee the danger of population growth or estimate the increasing consumption of petroleum products or the risks of nuclear energy used for peaceful purposes—as if these were merely *scientific* problems. Only the scientists, he says, could assess the concomitants and consequences of their own achievements. Solely because of this they would bear greater responsibility than others. Popper explicitly states that the accessibility of new knowledge creates new obligations. However, this is part of the special responsibility of the scientist within the framework of his or her *role obligation*. "Everyone has a special responsibility where (s)he has either special power or special knowledge." Popper would like to activate responsibility and its awareness by introducing a "promise" for students of applied natural sciences, to be oriented on the Hippocratic Oath of the medical profession (according to Weltfish, 1945). On the other hand, as previously mentioned, Lübbe (1980) judged that the scientist would be hopelessly overburdened with the assessment of and responsibility for the harmful side effects of scientific and technical progress. In view of the unforeseeable consequences of the extended scientific and technical possibilities for action and widespread outcomes, the concept of responsibility is therefore notoriously overstretched. Scientists and technicians *could* not bear the responsibility at all, because these decisions were *politically* responsible at the level of our public civic culture. It is probably not a question of assigning responsibility to one individual alone, but of (bearing) *co-responsibility*, of sharing responsibility. In this respect, are scientists to be absolved of any responsibility, given their particular individual position and also that within the system?"

Is not much to be expected from scientists for the socio-political and social aspects of future planning, and especially with regard to their lack of willingness to take responsibility, as some critics of society suspected? Did they even, as one critic wrote, "deliberately mislead" the population during the critical period of Chernobyl, and in some cases even "lied through their teeth" by providing falsely reassuring explanations? Could they (not) have overlooked the dangerous situation at all—in the double meaning of this expression? The physicist Wolf Häfele even saw Chernobyl not as a physical, but "*only*"(!) as a "semantic catastrophe."[26] How can he even maintain this after 10,000 deaths (after 10 years) and much larger numbers

---

[26] Lecture by Wolf Häfele on "Ausstieg aus der Kernenergie - wohin?" (Phasing out nuclear energy - where to?) at the University of Karlsruhe (winter semester 1986/87), as part of the introductory 'general studies' program.

of children who have died or grown up damaged by radiation? Not to mention the various forms of environmental contamination.[27]

In fact, Werner Heisenberg and Carl Friedrich von Weizsäcker had already discussed this issue following the report on the dropping of the Hiroshima bomb and had demanded that the individual scientists involved in research and development should take careful and conscientious account of the wider context. Weizsäcker, for example, said at the time that the American nuclear physicists had not made enough effort to gain political influence before the bomb was dropped; they had—as if they had great decision-making powers!—given the decision on the use of the atomic bomb out of their hands too early, especially since only scientists were capable, he thought, much like Popper later, of thinking objectively and dispassionately and, most importantly, in big picture terms. This optimism about the power of judgement—the better and special judgement of scientists—probably no longer seems generally justifiable today. Nevertheless, as already mentioned, it is precisely scientific associations such as the *Atomic Scientists of Chicago* (ASC) that have made a very responsible effort to address the moral problems of the accountability of research and its consequences. But unfortunately to no avail.

Applied scientific and technical developments, for example the development of the combustion engine or the production of dynamite or nuclear energy, of course, usually have the ambivalence of a positive and a destructive usability for society or mankind (now one talks of "dual use" technologies). Moreover, especially in areas such as genetic engineering and genetic biology (where basic research and the further development of technology are particularly closely intertwined and, as mentioned, merge seamlessly), basic research and technical development can no longer be separated as smoothly and easily as the idealized distinction between the "discoverer" and the "inventor" assumes.

Edward Teller, at any rate, was later aware of this role; only that he always withdrew into the role of the neutral expert who was fascinated by such a technically "sweet" project, as Robert Oppenheimer, the so-called father of the atomic bomb in the Manhattan Engineer District Project, had put it. It seems to me that many such statements are still too bound to the traditional individualistic concept of sole individual responsibility for the cause.

From the point of view of the aforementioned *extended* responsibility of man according to Hans Jonas (1979) and in the light of the above-mentioned divisibility of co-responsibility, one could speak in a more differentiated way of co-responsibility without attributing to the scientists and especially to the individual researcher a total sole responsibility. The extended responsibility and the implied openness and possibility of participation in view of the Faustian Pact for scientific and technological progress, which was once entered into and is no longer easy to revoke, is indeed more important than a retroactive individual moral responsibility for the sole cause of basic research projects, which can hardly ever be ascribed. It is important to make

---

[27] The same applies to the nuclear catastrophe at Kyschtym (Mayak) in 1957, which only became more precisely known after the fall of communism, and which released about 20 times as much radioactivity as in Chernobyl!

scientists, especially younger ones and students, aware of such extended and shared responsibility.

I would like to round off what has been said with the demands of a late university colleague, a physicist who was also President of the European Physics Society: In 1995, solid state physicist Werner Buckel had said in a lecture on the occasion of the 50th anniversary of the first experimental atomic bomb explosion at Trinity Site in New Mexico that[28] "in view of the many risks that can arise from scientific results"—nuclear research being only one example—it can no longer be said that: "The scientist provides only new insights. What is done with it is not his concern." "This line of argumentation," said Buckel, "must finally come to an end." This assertion is "not tenable and it is dishonest," if only because all scientists are very willing to take responsibility for positive developments from their results. But: "One cannot know what one will find. So, apart from a few examples, forbidding and demonizing scientific research cannot be the means to save humanity from perhaps bad developments. All research would then have to be stopped. Nobody can seriously want this, because it would deprive humanity of any chance of solving emerging problems."

"It is my firm conviction," continued Buckel, "that there is only one path that we should consciously take: We must try to achieve a responsible approach to the results of science. The scientists have a major task in this respect. They are in a better position than anyone else to foresee the consequences of their research results. They have to face this task and they have to say relentlessly what they can foresee as a possibility." Elsewhere in the same paper, he noted that the "attempts" to make completely irresponsible research in "high-risk areas" understandable to the "educated layman" "often have the character of defensive speeches": "One wants to convince the listener of something and to do so one chooses suitable arguments, which are certainly all correct, but not the full truth. The public is very sensitive to this." (I remember the word of a former German Chancellor who once drove into a journalist's parade: "That may be true, but it is not the truth").

"What we need," says Buckel, "are scientists who can point out all as yet conceivable[29] consequences, regardless of whether or not this suits the donor or any strong interest groups. He then also pays tribute to the Göttingen Declaration of the eighteen German nuclear scientists of 1957,[30] and believes that the refusal to cooperate in equipping the Bundeswehr with nuclear weapons was "responsible action in the best sense". He concludes with a few noteworthy demands:

1. "Scientists must not be bought. They may not make their scientific statements for or against any interest groups and possibly receive a particularly high fee for doing so." A high-ranking politician, reported Buckel, had once said to him publicly: "It's clear: I get a positive report for everything. The only question is how

---

[28] This event was held on 15 July 1995 under the title "Science in Responsibility" on the occasion of the 50th anniversary of the first nuclear test explosion (on 15 July 1945) by the German Physical Society, the Association of German Scientists, the Natural Scientists' Initiative "Verantwortung für den Frieden" (Responsibility for Peace) and the Göttingen Scientists' Association.

[29] "*All* conceivable"? This is an unfulfillable demand.

[30] See footnote 24.

much I'm willing to pay." Buckel's comment: "This is a scathing judgement on the morals of some scientists." (Just the *scientist*? I don't think so…)

2. "Scientists should strive to anticipate the (possible) consequences of their work (if possible). This takes effort because you have to obtain knowledge outside of your field."

3. "Scientists should relentlessly disclose what negative consequences their results can have"—in addition to the positive ones. "This would enable us to identify and avoid these consequences at an early stage. Research should not be "forbidden," but "learn to master its results. Research is crucial to solving our future problems."

4. "This behavior of scientists requires a certain change in the consciousness of our society. It must be recognized as a value in society if scientists behave responsibly by identifying potential risks at an early stage." (Wishful thinking?)

5. He then demands that when scientists speak as such, they should "put aside their personal opinions." But as citizens, they could have an opinion," they should not only "be allowed" to have one, but also "should have" one—an evaluative opinion, which just "does not need to be scientifically founded," cannot be scientific. I think a little further reaching: that scientists should certainly also express their personal opinions in the context of the debate on applications of research and the public discussion of science, but these personal opinions should be *labelled* as such.

Do Ethics Committees and a Scientific Oath Solve the Moral Problems?

One has often referred in particular to scientific ethics committees, which should be used not only in medicine but for *all* sciences. It seems doubtful to me, however, whether a permanent ethics committee—which would be concerned with the investigation and assessment of the ethical, social, and legal implications of basic research, and with progress not only in biomedical research, but in technology and in science in general—would be the appropriate institution to steer science, even if this committee were interdisciplinary and broadly based. Obermeier (1979), who suggested this, said that it was long overdue to regulate science before the permanent innovations and progress overwhelmed us. However, this would probably also assume an unrealistic predictability of scientific discoveries and their consequences. The super-experts, the super commission, would be institutionalized in this way. But they do not exist, cannot exist. It would indeed be absolutely overstretched. Even though ethics committees may be useful in biomedical and pharmacological research as well as in all human experiments for control purposes (because here people are directly and assessably subjected to the risk of the quite precisely known or, if possible, specified experiment),[31] a comprehensive commission is likely to find itself just as overwhelmed with the task of dealing with all the overarching problems of basic research as the individual scientist.

There is also useful commission work on individual questions, on concrete data definitions, which is undoubtedly very important and detailed. One thinks, for example, of the German Technical Instructions on Air Quality Control (Technische Anleitung zur Reinhaltung der Luft), which is also the result of careful commission work. In decision-making commissions of this kind, the scientists also assume quasi-legislative functions that fill out the framework guidelines of the laws, and this seems to be a very important transfer of overall responsibility today.

Some people, such as the retired biologist Hans Mohr, are apparently of the opinion that, ethically speaking, all this does not achieve anything (Mohr, 1979). The ethical commission solution could not work because science can only be morally judged and actually hardly ever be morally regulated in a way that is not influenced by political and social factors. Only the "ethos of science" functions for regulation, not the ethics of science. Otherwise, Mohr wrote by the end of the 1970s, no ethical uniformity among scientists could be achieved, as in life in general, nor could any "oath" be used to resolve honest differences of opinion and the legitimate pluralism of the scientific community on political issues. Politically, humanity is not a unity and cannot be brought to such a unity. But ethics is not only politics; and I do not believe that as an ethicist of science like Mohr has done his duty here. I think that he jumped to the extreme here, threw in the towel too quickly. In fact, humanity *must* come to a minimal consensus on survival, this must be a demand, a postulate of ethics: This is the only way to avoid a world catastrophe. It *must be* avoided. But even a "*Fiat moralitas, pereat mundus*" (morality must happen, even if the world may end) must not be a maxim. Incidentally, there are also certain fundamental convictions about the value of human life and its worthiness of preservation which are common to all cultures and societies and on which one can build.

---

[31] The convening of ethics commissions for the preliminary examination of all human experiments that may involve risks of harm is undoubtedly good, but practice is still controversial: some authors doubt the effectiveness and controllability of the commission; some fear the bureaucratic restrictions and requirements for research. A legal—at least something resembling "professional ethics"—regulation should also ensure the independence of the control; that seems indispensable and is widely supported, but the materialization of the well-taken idea was all too often given up or faltered due to increased bureaucratic cumbersomeness of the application, to industrial or political interests, checks and control procedures. For ethical reasons, however, such restrictions should be accepted for the sake of the people and ecosystems concerned.

The Hippocratic Oath is good as an idea, of course, but remains problematic because it has low effective-ness, (too) low controllability and enforceability.[32] It is good as an idea, but has a low effectiveness, (too) low controllability and enforceability. It does not take enough real political, practical action. It is at best ideal-typical. The idea of the Hippocratic Oath among scientists is not so absurd in human experiments and in research directly related to the experimental process. For the application of the results of completed research, other regulations should probably be used—strictly understood, more similar to the Hippocratic Oath (which is also primarily about the application of scientific knowledge or medical art in therapy). A certain co-responsibility of the scientist providing the procedures can be given on a case-by-case basis (particularly evident in the negative case: the scientific developer of napalm, Fieser, admittedly rejected any ethical co-responsibility just like Teller, the so-called "father of the hydrogen bomb"!). As said, the crux of the oath of science, which is analogous to the Hippocratic Oath of science, remains the low effectiveness, controllability, and enforceability. It is precisely an idea that is too general and abstract, too readily acceptable, and not concrete enough to be able to solve the ethical problems of research realistically.

The problem of ethical and legal control cannot be solved by the oath alone, especially since the career system of scientists has, in a certain sense, a built-in tendency in the opposite direction, namely incentives to violate ethical standards. A study of American medical researchers by Bernhard Barber (1976) showed that ambitious, upwards striving, and less successful scientists in particular tend to push

---

[32] Probably one of the first examples of proposed oath formulations for natural scientists came from Gene Weltfish (1945): "I pledge that I will use my knowledge for the good of humanity and against the destructive forces of the world and the ruthless intent of men; and that I will work together with my fellow scientists of whatever nation, creed or color, for these, our common ends." Newer versions include the following : Buenos Aires 1988 (International Symposium on Scientists, Peace and Disarmament): "Aware that, in the absence of ethical control, science and its products can damage society and its future, I pledge that my own scientific capabilities will never be employed merely for renumeration or prestige or an instruction of employers or political leaders only, but solely on my personal belief and social responsibility—based on my own knowledge and on consideration of the circumstances and possible consequences of my work—that the scientific or technical research I undertake is truly in the best interest of society and peace." Authors of the Institute for Social Inventions (n.d.; cf. also Lenk, 1991) in London proposed a Hippocratic Oath for Scientists, Engineers and Technologists: "I solemnly pledge myself to consecrate my life to the service of humanity; I will give to my teachers the respect and gratitude which is their due; I will practise my profession with conscience and dignity; The well-being of humanity will be my first consideration; I will maintain, by all the means in my power, the honour and the noble traditions of my profession; I will look on my colleagues as on my own family; I will not permit considerations of religion, nationality, race, politics or social standing to intervene between my work and my duty to humanity; I will maintain the utmost respect for human life from its beginning even under threat; I will abstain from whatever is deleterious and mischievous; I will not use my knowledge contrary to the laws of humanity; I make these promises solemnly, freely and upon my honor." In 1999, the Nobel Peace Prize winner (together with the Pugwash Conferences he chaired) Joseph Rotblat proposed the following version: "I promise to work for a better world, where science and technology are used in socially responsible ways. I will not use my education for any purpose intended to harm human beings or the environment. Throughout my career, I will consider the ethical implications of my work before I take action. While the demands placed upon me may be great, I sign this declaration because I recognize that individual responsibility is the first step on the path to peace" (Rotblat, 1999).

ethical considerations completely aside in human experimentation and, in the interest of their own scientific careers, to produce exciting or sensational experiments and results quite quickly. This is of course a dangerous development. In a certain sense, ethics committees can certainly introduce a code of ethics here, a limitation on ethically unacceptable human experiments that has indeed occurred. In any case, mere career considerations should not exacerbate the dilemma of human experiments. Checks do indeed appear necessary. However, making them effective with minimal impediment to research is also a valuable research postulate—and presents a very difficult ethical demarcation problem.

One example is the (quasi-sporting) the competition to fully sequence the human genome. Such conflicts are built into any dynamic research in particular, and often seem to be an indispensable motivational force. Ideally, such motivations should not be at the expense of human test subjects, especially not the individuals involved in invasive physical or psychological experiments. On the subjects' side, ethics is certainly first and foremost the ethics of the individual, usually aimed at ensuring their overriding personal integrity.

In the interests of many of those affected, however, science should not be unnecessarily hindered or prevented. So far, there are no patent remedies for all-round solutions to such conflicts. One must do everything possible to raise the awareness of conflict and support scientists, such that they are not compelled to decide unilaterally in the case of application, meaning not to blindly follow one's own career interests and not to suppress ethical considerations.[33]

Co-responsibility Without Sole Responsibility

The responsibility of the researcher in science and technology is indeed a special case of role-specific and moral responsibility in a strategic position. Consideration of the aforementioned fiduciary responsibility for prevention and protection is required wherever harmful effects can be estimated and averted, e.g., in directly application-oriented scientific and technical projects. A personal co-responsibility may be given on a case-by-case basis, but a general strict or even sole causal responsibility of scientists and technicians does not exist in any case, especially in basic research, in view of the ambivalence and collective origin of research results. This means in fact: *co-responsibility to be* differentiated and concretized in more detail *without sole personal responsibility*. We must find such a middle solution. All the more important is the preventive responsibility, the responsibility to prevent destruction and permanent damage in advance. In view of the dynamics of development

---

[33] Unethical research corresponds roughly to covert foul play or doping in elite sport. Not only is it unfair, as in sports competition, to gain and use an advantage for oneself through some kind of rule violation, but it would also be unfair and unethical to damage or harm others: in sports, mostly one's opponents, or in research one's test subject or uninvolved but affected persons. Perhaps one should not be too "sports-like" with research in particular. But this remains an idle call in a time of intensified and still increasing competition for research results and positions—as well as for researchers' reputation and qualification.

and the difficulties of orientation and evaluation in this entire problem area, it seems to me that the only realistic way to adequately turn to the ethical challenges of the future is to promote moral awareness in questions of the ethics of science among all scientists, if possible, and especially among prospective young scientists, and to discuss the interrelationships related to individual research projects, especially on the basis of concrete case studies.

A change has taken place in medical research, especially of course in the ethical and legal debates on genetic engineering and more recently in stem cell research. In other applied sciences they lag even further behind. Ethics should therefore not only be demanded and promoted as a school subject, but should also be developed as a scientific–ethical awareness subject for moral conscience training in the field of research, especially in the education geared towards it. There is moral importance in appropriately training the consciousness of prospective scientists and technicians. Only if this (moral awareness) is widely stimulated and practiced will it be possible to recognize more precisely the extended co-responsibility, the division of responsibility without deduction of responsibility and without attributing sole responsibility to the scientists themselves, and will it be possible for them to handle it.

Sensitive Co-responsibility: Comments on "Scientific Freedom and Scientific Responsibility"

My following considerations responded to the theses on responsibility in science conveyed in the publication "Scientific Freedom and Scientific Responsibility: Recommendations for Handling Security-Relevant Research" published by the German Research Foundation (DFG) in cooperation with Leopoldina, the German National Academy of Sciences (DFG & Leopoldina, 2014).

Problems of responsibility in the sciences become all the more urgent the more that scientific knowledge, skills, and technical as well as political or institutional power grow, and the more the technical world is shaped by them. Power, skill, *and* knowledge do hold someone responsible. In the systems-connected world, the allocation of responsibility to individuals alone is no longer sufficient.

Total neutrality of the scientist and science as an institution/professional association is just as unrealistic as a sole responsibility of the individual scientist and technical researcher would be. Their informed and sensitive *co-responsibility* is necessary, especially in security-related research. *System responsibilities* are easy to demand, but very difficult to deal with in practical–operational terms.

Analytically one should, as far as possible, continue to distinguish between the model poles of "pure basic research" and "technical application," between "discovery" and "development." However, reality today is mostly concretized in intermediate types, mixed types, e.g., in application-oriented basic research or fundamental ("knowledge-oriented") purpose-oriented research or in the (e.g., information and biotechnological) development of purpose-oriented methods (basic) research.

Participation models should be developed in order to make more comprehensible, operational, and tangible not only the *internal* responsibility of the guilds ("the ethos of science") but also the *external* co-responsibility of scientists and technicians towards society and mankind. The idea of sharing responsibility according to centrality and influence via power, strategic position in research, and decision-making processes—and through knowledge in this regard—needs to be elaborated more precisely. Institutional procedural regulations for assessment and possible sanctions should be developed (protection of particularly morally acting experts, e.g., *"whistle blowers"*, awards, opportunities for discussion for training and orientation purposes, hearings, committees, etc.), without simply subjecting *everything* to legal regulation or an ethical paternalism of everything (e.g., through blanket "can" formulations, see below) and/or through bureaucratic supercommissions: Ethics goes beyond mere legal regulations. It is important, however, to involve the scientists themselves in the fundamental interdisciplinary discussion that goes beyond science itself (This took many decades, namely until the so-called *codes of ethics,* already in the USA, e.g., IEEE, APA, as well as ethics committees in German science, technology, medicine, etc.).

Different types and forms of responsibilities (see above) are to be distinguished as "analytically clean" models (as "ideal types"): They can usually overlap or conflict with each other. In order to clearly identify and prepare the resolution or mitigation of conflicts of responsibility, it is urgent to examine them more closely.

Priority rules of (un)responsibilities as orienting guidelines (e.g.: direct and moral responsibility takes precedence over indirect and role responsibility) are to be more precisely drafted, elaborated, and reviewed in (not only subject-related) expert groups of scientific and technical associations, also through public discussion, but including practice-oriented philosophical–ethical and social science analysis.

This, and also the consideration of the internationally quite differentiated debate (and the national preparatory work) is still somewhat lacking, especially since the new "Recommendations" of the DFG and the Leopoldina, as important as the (rather belated)start is.

A few more details on the "Recommendations" themselves: In these it is rightly emphasized that the specific "guild-like" internal responsibility—and also the universal moral (general–ethical) one—goes or "can" go beyond the purely legal "obligation." New considerations include the binding requirement of transparency (with justified exceptions), the emphasis on the *"dual-use"* problem of protecting constitutional "goods" and "values", the justified weighing and monitoring of risks of damage and a "should be" obligation of prior and accompanying consideration of problems, of consequences, implementation, controllability, and further publication, as well as an institutional obligation to raise legal and ethical awareness of abuses, including the problems of whistle-blowers and their protection.

It must be criticized, however, that apart from the twice mentioned (but not specified in any way) "special responsibility" of the scientist and that the "primary goal" is "to carry out and communicate research in a responsible way", no differentiating statements at all can be found about the different types and kinds of responsibility of the scientist and the scientific institutions: Vague and meaningless or ultimately

non-binding formulations such as constant "can" statements (instead of the normative "should" or equally rare, but here too sometimes misleading, "must" standards)
are editorially objectionable. More importantly, no distinction whatsoever can be
made between the very different types and nature of responsibilities (apart from
personal, institutional, and legal), let alone certain rules or orienting guidelines for
dealing with (completely unnamed, but nevertheless typical) *conflicts of responsibility*, for example in the sense of (ethically or constitutionally based) priority rules
(existing analyses have simply not been taken note of). Despite the reference to
"concrete measures," the latter are not mentioned (except for the reference to awareness training). However, it is stated that the "measures" (which are not at all specified in the recommendation) "must not be permitted to inappropriately hinder
research and are subject to feasibility and proportionality." (what does this mean
and who would decree that?).

All in all, these (without the participation of analytical–philosophical or ethical
experts) are very amateurishly "hand-knitted" "recommendations" without a truly
theoretical–analytical clarification and practically guiding function.

Theses on Responsibility in Science

Finally, I would like to mention fifteen theses on responsibility in science,[34] which
of course still need to be explained and supplemented in detail:

1. Problems of responsibility become more urgent the more scientific knowledge
   and technical power grows and the more the technical world is shaped. Power
   and knowledge make someone (co-)responsible.
2. Total neutrality of the scientist and science as an institution/professional association is as unrealistic as a sole responsibility of the scientist and technician
   would be.
3. Analytically speaking one should, as far as possible, continue to distinguish
   between the model poles of "pure basic research" and "technical application,"
   between "discovery" and "development." Otto Hahn could not be held responsible for the development of the atomic bomb, but Edward Teller was partly
   responsible for the H-bomb. However, reality today is mostly concretized in
   intermediate types, mixed types, e.g., in application-oriented basic research or
   basically purpose-oriented research or in the (e.g., information) technological
   developments of purpose-oriented methods (basic principles).
4. In some new fields of research—for example in scientific computer science and
   information technology, but especially in biotechnology, for example in genetic
   biology and genetic engineering—basic research and possible applications are
   so closely linked, even interwoven, that there is often no real separation between

---

[34] Based on old theses from 1996 to 1998, here in selection, edited and supplemented, presented
with the comments from a colloquium at the Karlsruhe Institute of Technology, 2015.

experimental research, applied science and technology or scientific technology. Research and development results can sometimes lead almost immediately and very quickly to possible applications. The *explosiveness* of external responsibility problems in biomedical research, biotechnology, for example, is penetrating the sciences, and is intensifying the problem of responsibility even for the basic researcher.

5. Participation models must be developed in order to make the external co-responsibility of scientists and technicians towards society and "humanity" and to make group and co-responsibility more comprehensible, operational, and tangible. The idea of sharing responsibility according to centrality and influence through power and knowledge must be elaborated. Institutional procedural regulations for assessment and possible sanctions should be developed (protection of particularly morally acting experts, awards, discussion opportunities for training and orientation purposes, hearings, etc.), without simply subjecting everything to legal regulation or the ethicization of everything and everyone by bureaucratic supercommissions: Morality goes beyond legal regulations (Germans, in particular, like to suppress this insight). It is important to involve scientists themselves in the interdisciplinary discussion which goes beyond science itself.

6. Different kinds and types of responsibilities must be distinguished analytically (see above). They can and do overlap each other in social reality, or more often, yes, typically conflict with each other. In order to clearly identify—and prepare the solution to—conflicts of responsibility, they need to be examined more closely.

7. Practical relevance and empirical studies of observations and experiences as well as social-psychological, group-dynamic experiments should be initiated in a targeted manner and should take place in a problem-oriented manner, the results of which can be easily grasped.

8. Priority regulations (see above) of responsibilities (e.g., direct and moral responsibility takes precedence over indirect and role responsibility) are to be drafted and reviewed by means of possibly public discussion (e.g., also in and by commissions of inquiry) and academic or philosophical analysis.

9. In practical situations of conflicts of responsibility and conscience, the concrete human responsibility and conscience decision of the individual should be decisive, but this cannot be understood alone and isolated as the only foundation.

10. If conscience is indeed the conscious self-attribution, i.e., the explicit assumption and self-experience of responsibility, then a differentiated discussion and training of responsibility in the form of knowledge of different types of responsibility and the conflicts that frequently arise between them, as well as guidance and practical handling of the *combination of these different responsibilities in a concrete-human form, is at the same time also a differentiated cultivation of conscience.*

11. The social embedding and institutional design or standardization and orientation of the orientation of the variants of responsibility and conscience are necessary. Particularly in education and further training, there should also be training

in the culture of responsibility and conscience. The training of the individual conscience and especially the sensitive awareness and attention to different responsibilities and conflicts of responsibility is just as important in an increasingly complex society characterized by different loyalties as it is for the social subsystem of applied science or technology research.

12. Science and technology researchers and analytically trained moral philosophers with in-depth knowledge of scientific work, research, and innovation would have to cooperate and increasingly examine the finer practical structures of responsibility and relate their types, kinds and levels to each other and model them as realistically as possible.

13. In this way—and only in this way—can ethicists and moral philosophers also take on their own special *meta-responsibility of improving the conceptual methodology and social philosophy* with practical and truly beneficial prospects.

14. Practical and *concrete humanity* should always be a central guiding principle: *In dubio pro humanitate practica!*

15. As far as the ethical debate as a whole is concerned, in view of the challenges of applied science, research, and technology, we are unfortunately still almost at a beginning. It does not require a prophetic ability to put forward the thesis already mentioned in passing: We cannot afford today, and certainly not in the future, to neglect the urgent ethical problems of the applied sciences and in the world of technology (research) and business as we have done in the past.

We must differentiate the prophetic words of Marx: "The philosophers have only *interpreted* the world in different ways; but it is important to *change* it" in a *responsible and sustainable way.*[35]

References

Barber, B. (1976). The ethics of experimentation with human subjects. *Scientific American, 234*, 25–31.

Belsey, A. (1978). The moral responsibility of the scientist. *Philosophy, 53*, 113–118.

Belsey, A. (1979). Scientific research and morality. In *Contribution to the sixth international Congress on logic, methodology and philosophy of science* (pp. 211–215). Hanover.

Bodenheimer, E. (1980). *Philosophy of responsibility.* Rothman.

Born, M. (1969). Die Zerstörung der Ethik durch die Naturwissenschaften. In H. Kreuzer & W. Klein (Eds.), *Literarische und naturwissenschaftliche Intelligenz* (pp. 179–184). Klett.

Buckel, W. (1995). Wissenschaft in der Verantwortung. Lecture in Göttingen (15.7.1995).

Chain, E. (1970). Social responsibility and the scientist. *New Scientist, 48*, 166–170.

DFG & Leopoldina. (2014). *Recommendations for handling security-relevant research.* Bonn/Halle. Retrieved March 16, 2021, from https://www.dfg.de/download/pdf/dfg_im_profil/reden_stellungnahmen/2014/dfg-leopoldina_forschungsrisiken_de_en.pdf

French, P. A. (1984). *Collective and corporate responsibility.* Columbia University Press.

Hardin, G. (1968). The tragedy of the commons. *Science, 162*, 1243–1248.

---

[35] Karl Marx's eleventh Feuerbach thesis (translated). The original sentence ends after "change it."

Herneck, I. (1977). *Bahnbrecher des Atomzeitalters*. Der Morgen.

Herrmann, A. (1977). *Die Jahrhundertwissenschaft: Werner Heisenberg und die Physik seiner Zeit*. Deutsche Verlagsanstalt.

Herrmann, A. (1982). *Wie die Wissenschaft ihre Unschuld verlor*. Deutsche Verlagsanstalt.

Ingarden, R. (1970). *Über die Verantwortung*. Reclam.

Jonas, H. (1979). *Das Prinzip Verantwortung*. Suhrkamp (English 1984: The Imperative of Responsibility)

Ladd, J. (1990). *A comprehensive theory of moral responsibility*. Unpublished manuscript.

Lang, W. (1985). Responsibility and guilt as legal and moral concepts. *Archiv für Rechts- und Sozialphilosophie, 24*, 262–268.

Lenk, H. (1979). *Pragmatische Vernunft*. Reclam.

Lenk, H. (1982). *Zur Sozialphilosophie der Technik*. Suhrkamp.

Lenk, H. (Ed.). (1991). *Wissenschaft und Ethik*. Reclam.

Lenk, H. (1992). *Zwischen Wissenschaft und Ethik*. Suhrkamp.

Lenk, H. (1996). Zur Verantwortung des Forschers. *Jahrbuch Wissenschaft und Ethik, 1*, 29–71.

Lenk, H. (1997a). *Einführung in die angewandte Ethik*. Kohlhammer.

Lenk, H. (1997b). *Konkrete Humanität*. Suhrkamp.

Lenk, H. (2000). *Albert Schweitzer: Ethik als konkrete Humanität*. LIT.

Lenk, H. (2005). Responsibility: German perspectives. In C. Mitcham (Ed.), *Encyclopedia of science, technology, and ethics* (Vol. 3, pp. 1618–1623).

Lenk, H. (2006). *Verantwortung und Gewissen des Forschers*. Studienverlag.

Lenk, H. (2007). *Global technoscience and responsibility*. LIT.

Lenk, H. (2015). *Human-soziale Verantwortung*. Projekt-Verlag.

Lenk, H. (2019). *Not a long way to concrete humanity?* Projekt-Verlag.

Lenk, H., & Ropohl, G. (Eds.). (1987). *Technik und Ethik*. Reclam.

Lübbe, H. (1980). Wissenschaftsfeindschaft und Wissenschaftsmoral. In P. Labudde & M. Svilar (Eds.), *Wissenschaft und Verantwortung* (Lecture series at the University of Bern) (pp. 7–17).

Mackie, J. L. (1977). *Ethics*. Pelican Books.

Maring, M. (2001). *Kollektive und korporative Verantwortung*. LIT.

Mohr, H. (1979). The ethics of science. *Interdisciplinary Science Reviews, 4*, 45–53.

Mößbauer, R. (1994). *Interview with "Ventil"*. Universität Karlsruhe: Asta, no. 94.

Neumaier, O. (1986). Die Verantwortung im Umgang mit dem Begriff der Verantwortung. In O. Neumaier (Ed.), *Wissen und Gewissen* (pp. 213–228). VWGÖ Wien.

Neumaier, O. (2008). *Moralische Verantwortung*. Schöningh.

Obermeier, O. P. (1979). Darf der Mensch alles machen, was er kann? *Politische Studien, 30*, 565–574.

Popper, K. R. (1977). Die moralische Veantwortung des Wissenschaftlers. In K. Eichner & W. Habermehl (Eds.), *Probleme der Erklärung sozialen Verhaltens* (pp. 294–304).

Rotblat, J. (1999). A Hippocratic oath for scientists. *Science, 286*(5444), 1475.

Stoner, J. A. F. (1968). Risky and cautious shifts in group decisions. *Journal of Experimental Social Psychology, 4*, 442–459.

Ströker, E. (1986). Inwiefern fordern moderne Wissenschaft und Technik die Philosophische Ethik heraus? *Man and World, 19*, 179–202.

Teller, E. (1945, July 4). *Letter to Leo Szilard*. Retrieved March 9, 2021, from, https://www.atomicarchive.com/resources/documents/manhattan-project/teller-petition-response.html

The Institute for Social Inventions. (n.d.). *The Hippocratic oath for scientists, engineers and technologists*. Retrieved March 13, 2021, from, https://explore.scimednet.org/wp-content/uploads/2016/05/hippocratic-oath.pdf

Weltfish, G. (1945, September 24). Scientists' oath. In: *Science*.

Werhane, P. H. (1985). *Persons, rights, and corporations*. Prentice-Hall.

Chapter 3
The Principle of Responsive Adjustment in Corporate Moral Responsibility: The Crash on Mount Erebus

Peter A. French

Abstract The tragic crash of Air New Zealand flight TE-901 into Mt. Erebus, Antarctica provides a fascinating case for the exploration of the notion of corporate moral responsibility. A principle of accountability that has Aristotelian roots and is significantly different from the usual strict intentional action principles is examined and defined. That principle maintains that a person can be held morally accountable for previous non-intentional behavior that had harmful effects if the person does not subsequently take corrective measures to adjust their behavior so as not to produce repetitions. This principle is then applied to the Mt. Erebus disaster.

1. On the morning of November 28, 1979 flight TE-901, a McDonnell Douglas DC-10 operated by Air New Zealand Ltd., took off from Auckland, New Zealand on a sightseeing passenger flight over a portion of Antarctica. The following are paragraphs from the official Report of the Royal Commission that inquired into the events surrounding that flight.[1]

   [#12] The personnel at McMurdo Station and Scott Base were expecting the arrival of an Air New Zealand DC-10 aircraft carrying sightseeing passengers. The flight plan radioed to McMurdo from Auckland had named the pilot in command as Captain Collins. ...It was expected that the DC-10 would fly down McMurdo Sound. ...The aircraft would come in

---

---

[1] *Report of the Royal Commission to Inquire into the Crash on Mount Erebus, Antarctica, of a DC-10 Aircraft Operated by Air New Zealand Limited, 1981.* Presented to the House of Representatives by Command of His Excellency the Governor-General. Hereafter, references to this report will be made in the text by citing the paragraph numbers of the Report.

P. A. French (✉)
Arizona State University (Emeritus), Tempe, AZ, USA
e-mail: Peter.French@asu.edu

© The Author(s) 2022
H. A. Mieg (ed.), *The Responsibility of Science*, Studies in History and Philosophy of Science 57, https://doi.org/10.1007/978-3-030-91597-1_3

from the north and in the vicinity of Ross Island would descend to a low level so as to afford the passengers… sightseeing. … The aircraft would probably fly down the Sound at an altitude of somewhere between 1500 feet and 3000 feet, … a perfectly safe altitude at which to fly over flat ground in clear weather, and the cause of no concern to the United States Air Traffic Control.

[#13]…When the DC-10 was about 140 miles out from McMurdo, Mac Centre transmitted a weather forecast… to the effect that there was a low overcast over Ross Island and the McMurdo area. …Mac Centre suggested that once the aircraft was within 40 miles of McMurdo Station, the entrance of McMurdo Sound, it would be picked up by radar and its descent through cloud guided down to an altitude of 1500 feet. This suggestion was accepted by the air crew. At 1500 feet, under the cloud layer in the McMurdo area, visibility would be unlimited in all directions… .

[#15] By 12:35 p.m. it was confirmed between Mac Centre and the DC-10 that the aircraft was now descending to 10,000 feet and was requesting a radar let-down through cloud. The request was accepted by Mac Centre… .

[#16] At 12.42 p.m. the aircraft informed Mac Centre that it was flying VMC (visual meteorological conditions) and that it would proceed visually to McMurdo. This message indicated that the aircraft had found an area free of cloud through which it would descend before leveling out at an altitude less than the cloud base. Thus the aircraft would be approaching lower than the cloud layer, in clear air at an altitude of about 2,000 feet… .

[#17] There followed further transmissions between the aircraft and Mac Centre and then at 12:45 p.m. the aircraft advised Mac Centre that it was now flying at 6,000 feet in the course of descending to 2,000 feet and that it was still flying VMC… . This was the last transmission from the DC-10.

[#22] The United States Navy sent out aircraft on intensive searches and ultimately, after several hours, the reason for the long radio silence from the aircraft was discovered. A United States Navy aircraft found the wreckage of the DC-10 on the northern slopes of Mount Erebus at a point about 1,500 feet above sea level. The aircraft had been carrying 20 crew and 237 passengers. There were no survivors.

2. Those familiar with the unhappy history of the DC-10 naturally would suspect that another engineering fault had been responsible for this disaster. The Mount Erebus crash, however, was not the result of any airplane engineering errors or faults. This DC-10 performed perfectly. The simple fact is that the plane was flown in broad daylight in 'clear air' at approximately 2000 feet above sea level directly into the side of a 12,000-foot mountain. 'Simple facts' in the case of airline disasters are, however, notoriously complex. Photographs taken by the ill-fated passengers indicate that just seconds before the crash the view from the plane was unobstructed for many miles. The report notes that:

It followed (from the photographic evidence) that as the aircraft had approached Mount Erebus it was flying in skies in which there was perfect visibility for at least 23 miles. It was also apparent that the aircraft had been flying well under the cloud base when it collided with the mountain. [#28]

Air New Zealand Ltd. released a statement proclaiming that the cause of the crash had been 'pilot error'. The airline maintained that Captain Collins had become disoriented or confused or distracted and had flown TE-901 directly into the mountain. They added that he had no business flying that airplane at 2000 feet. The flight, they said, should have remained at no less than 14,000 feet, a claim that the Royal Commission Report disputes by pointing out that the very nature of the flight necessitated relatively low altitudes. At 14,000 feet the passengers would have seen nothing but clouds. They would certainly not have witnessed the advertised splendor of Antarctica. And, importantly, clearance to fly at 1500 feet down McMurdo Sound was granted by the Mac Centre controller and was generally regarded as a safe altitude for such flights over the flat terrain of the Sound.

The bald fact is that Captain Collins had flown far off course, some 27 miles, and that he was flying into Lewis Bay not the McMurdo Sound. Lewis Bay is a much smaller body of water than the Sound and from its shore on Ross Island rises Mount Erebus. Captain Collins was no fledgling pilot. He had a long and distinguished flying career. The 'pilot error' theory called for him to have made a monumental miscalculation as to his location and also to have failed somehow to see the impending doom of the slopes of Mount Erebus rising up directly in front of him.

3. The evidence collected by the Royal Commission established quite a different scenario than provided by the airline. In order to understand their version we need to take a brief digression to explain something about the navigational systems on the DC-10. DC-10s are navigated by a computer system known as the Inertial Navigation System (INS). DC-10s proceed to their destinations by a series of waypoints. The INS

   operates by typing into a computer system on the aircraft (the AINS) the latitude and longitude of each waypoint.... . Once this series of coordinates has been fed into the aircraft's computer, the aircraft will then fly its own course from one waypoint to another. [#31]

This flight path is known as the aircraft's 'nav track'.

   TE-901 was, its recovered black box revealed, flown on its programmed nav track almost the entire distance of the flight. It is also important to note that INS systems are widely regarded in the industry to be highly accurate. [#33]

How then did TE-901 end up in wreckage on the slopes of Mount Erebus when it should have been cruising at 2000 or 1500 feet above McMurdo Sound, passenger cameras clicking wildly at the spectacular vistas of Antarctica?

4. The primary cause of the disaster, the Royal Commission reported, was a direct result of the administrative and communication system of Air New Zealand Ltd. In order to see how they reached this conclusion it is necessary to briefly canvass some of the facts uncovered by the Commission.

Captain Collins and his First Officer attended a briefing on Antarctic flights nineteen days before their TE-901 assignment. At that meeting they were provided with printouts of the flight plan used by a sightseeing flight that had just returned from Antarctica. The coordinates on that flight plan, they were informed, would be the standard ones for all of the airline's sightseeing flights. The Commission learned that Collins had written the coordinates in his notebook and that on the night before TE-901 was to depart, he had plotted the track of his forthcoming flight on his atlas and a map. That track showed that the plane would head directly down McMurdo Sound with Mount Erebus approximately 27 miles to the east. The Report reads:

When the flight crew assembled on the morning of the flight and were handed the flight plan for 28 November 1979 extracted from the ground computer earlier in the morning and when the flight crew inserted into the computer on the aircraft the series of latitude and longitude coordinates on that flight plan they believed, in accordance with ordinary and standard practice, that they were inserting the longstanding coordinates. [#36]

The tragic fact was that they were doing nothing of the kind. A set of figures different than those provided at the briefing had been substituted and that change had the effect of moving the destination waypoint for McMurdo Sound 27 miles to the east, directly at Ross Island and collision with Mount Erebus (if the plane were flown at less than 12,450 feet). Collins, according to standard operating procedure, holding to his computerized nav track, had flown the plane directly into the mountain, in the mistaken belief the plane was heading down a wide and open expanse of water.

The computer coordinates were changed in the airline's Flight Operations Division. That Division is comprised of four subunits: Navigation Section, Computer Section, Flight Dispatch Section, and the RCU [Route Clearance Unit] Briefing system. Navigation Section directed the Computer Section to reprogram the computer after it had received a verbal direction to do so from Captain R.T. Johnson, operations manager for DC-10 aircraft at Air New Zealand. There is no record of any direction to Flight Dispatch Section to inform Captain Collins of the change nor any to indicate that the change had actually been made.

There are no records of any kind in Flight Operations Division. It is standard operating procedure in the airline that none of the decisions in the Division are committed to writing.

As it happens, the alteration of the nav track was a very reasonable thing to do. Captain Simpson, who piloted the Antarctic flight prior to TE-901, had noticed that the programmed waypoint for McMurdo was approximately 27 miles to the west of the tactical air navigation system (TACAN) of McMurdo Station. [#242] (The TACAN allows a pilot to determine his distance from it and hence is a useful navigational aid.) Simpson informed Captain Johnson that crews of future flights should be aware of this discrepancy, but he now claims that he did not suggest that a change in the flight plan be made. [# 243–245] Johnson, however, believed that Simpson had told him there was a serious error in the flight plan and that the relevant waypoint ought to coincide with the location of the TACAN. Johnson testified that he believed that when the coordinates were changed in the computer there would be no need to inform Captain Collins because the alteration amounted to only a 2.1-mile movement of the final waypoint. In this calculation he was woefully wrong.

The Commission explained the nav track change in a radically different way. Frankly, they did not accept Captain Johnson's claim that Simpson's information had been misinterpreted. Instead, they argue:

Both Captain Johnson and the Navigation Section knew quite well that the McMurdo waypoint lay 27 miles to the west of the TACAN and that since this track had not officially been approved by the Civil Aviation Division, it should therefore be realigned with the TACAN and then someone forgot to ensure that Captain Collins was told of the change. Such an interpretation means that the evidence as to the alleged belief of a displacement of only 2.1 miles is untrue. [#245]

The nav track alteration, the pilot's reliance on the AINS combined with the practice of descending to a low altitude to allow better views, indeed any view at all, took TE-901 directly into Mount Erebus. The Commission held that it would not be just to hold Captain Johnson fully responsible for the disaster. Although Captain Johnson probably ordered the alterations in the computer to put the Antarctic flights on a course that provided the pilots with a better set of navigational checks than did the path down McMurdo Sound, it is clear that he had no knowledge of the other key factor that brought TE-901 to disaster.

5. Certainly an experienced pilot, even if he believes himself to be flying over a relatively wide body of water, would quickly correct his descent from 16,000 feet when he saw a 12,450-foot obstacle in his direct path. Nav track or no nav track, an experienced pilot can be expected to take manual control and ascend to a safe altitude, fly over Mount Erebus and circle east to the Sound. Why had Captain Collins not corrected his descent when the face of Mount Erebus loomed up before him?

Although the answer is quite straightforward, it took some detective work for the Commission to discover it. They learned that in 'clear air' in Antarctica a pilot can experience a visual difficulty called a 'whiteout'. The whiteout is an atmospheric effect that produces a loss of depth perception, flattening out even mountainous terrain.

> Only two conditions are necessary to produce whiteout: a diffuse, shadowless illumination and a monocoloured white surface... . The condition may occur in a crystal-clear atmosphere or under a cloud ceiling with ample comfortable light. [#165]

It is likely that TE-901's pilot and flight crew experienced a whiteout. The Commission itself experienced the phenomena while on an investigative trip to the crash site. The crew of TE-901 never saw Mount Erebus or rather they never recognized what they saw as a 12,450-foot mountain. Their belief that they were flying over the waters of McMurdo Sound was apparently confirmed for them by what they thought they saw out of their windows.

6. The Commission listed ten factors or circumstances which they believed contributed to the crash. They are as follows:

(1) Captain Collins had complete reliance upon the accuracy of the navigation system of his aircraft... .
(2) There was not supplied to Captain Collins, either in the RCU briefing or on the morning of the flight, any topographical map upon which had been drawn the track along which the computer system would navigate the aircraft.
(3) Captain Collins plotted the nav track himself on the night before the flight on a map and upon an atlas.
(4) The direction of the last leg of the flight path to be programmed into the aircraft's computer was changed about six hours before the flight departed.
(5) Neither Captain Collins nor any member of his crew was told of the alteration which had been made to the computer track.

(6) Checks made in flight... demonstrated that the AINS was operating with its customary extreme accuracy... .

(7) McMurdo Air Traffic Control believed that the destination waypoint of the aircraft was 27 miles west of the McMurdo Station... .

(8) Mac Centre invited the aircraft to descend to 1,500 feet in McMurdo Sound for the reason that visibility at that altitude was 40 miles or more.

(9) Captain Collins accepted this invitation and made the decision to descend to that altitude.

(10) The nature of the cloud base in the area and the unrelieved whiteness of the snow-covered terrain beneath the overcast combined to produce the whiteout visual illusion. [#387]

The Commission maintains that had any of these factors (or what Mackie would have called INUS conditions)[2] not been present, TE-901 would not have crashed into Mount Erebus.

Despite the fact that the ten factors in this case were all necessary, though individually insufficient, parts of the set that though not necessary were sufficient to result in the crash, the Commission felt it was able to identify two acts, (2) and (5), both omissions on the part of the airline, upon which to found or justify ascriptions of moral responsibility for the crash to the corporation. It was concluded that, "The dominant cause of the disaster was the act of the airline in changing the computer track of the aircraft without telling the aircrew". [#392] From the point of view of the Commission, factor (5) was the INUS condition with a difference.

7. This lengthy, though sketchy version of the details of this case is justified if we want to understand what the Commission thinks their identification of the 'dominant and effective cause' of the Mount Erebus crash entails. Why had they not simply laid the moral responsibility for the disaster at Captain Johnson's doorstep? He, as the Simpson testimony seems to reveal, ordered the change in the computer track and failed to ensure that the information about the new track was communicated to the on-board captain. The Commission, however, refused to lay responsibility squarely on Johnson's shoulders. It concluded its report by maintaining that the crucial mistakes regarding the reprogramming and the failure to inform are "directly attributable, not so much to the persons who made (them), but to the incompetent administrative airline procedures which made the mistake(s) possible". [#393] In other words, the Commission concluded that the cause of the Mount Erebus disaster that founds an ascription of moral responsibility has an organizational nexus, and that it would be unjust to attribute the crucial failures to individuals within the company. In an important and irreducible way, they held that the standard operating procedures of Air New Zealand were causally responsible for the crash. Is this justifiable? And, more impor-

---

[2] J. L. Mackie, 'Causes and Conditions', *American Philosophical Quarterly* 2, No. 4 (1965), pp. 245ff; Here, Mackie defined INUS as an: Insufficient but Necessary part of a condition which is itself Unnecessary but Sufficient.

tantly, can it be translated to an ascription of moral responsibility to the corporation?

The Commission's investigation focused on the standard operating procedures of the Flight Operations Division of Air New Zealand. Four major defects in the administrative structure of the Division were identified. The flaws described by the Commission were:

1. Operational pilots held executive positions in the Flight Operations Division;
2. None of those operational pilots had been given a training course in administrative management;
3. There exist no written directives from Flight Operations Division that spell out the duties and administrative responsibilities of the executive pilots;
4. There exist no written directives in Flight Operations that specify the way various duties are to be performed, e.g., there exists no written directive that specifies the steps to be taken by the various sub-sections when adjustments are made to flight plans or changes are made in navigational procedures [see #365].

The structure and practices of Flight Operations depended upon a traditional of verbal communication. There is, in fact, a general policy in Air New Zealand against written memoranda. The Chief Executive Officer (CEO) of the airline testified before the Commission that he had always "controlled the airline on a verbal basis". [#366] Communications regarding inter- or intra-departmental business were almost never written. The fatal alteration in the coordinates was, according to standard practice, verbally directed to the Navigational Section by Captain Johnson. If any confirmational or informational messages passed from one subsection of Flight Operations to another or to the flight crew, they would have been verbal.

It might be argued that the CEO or the Board of Directors should have expected their verbal communication policy to result in serious breakdowns. The fact that active pilots also wore administrative hats and hence, on many occasions, had split duties and subsequently split attentions, should have also alerted some concern. A pilot with a number of his own flight problems on his mind could easily forget, misunderstand, or confuse information passed to him verbally, especially when that information concerned flights other than those he regularly flew. The Commission, however, did not make such an argument.

The Commission holds the view that the Board and the CEO could not be expected to investigate the daily administration of Flight Operations and that there was a long history of excellent, safe airline service at Air New Zealand to justify whatever smugness might have been manifested regarding the lack of written documentation of administrative communications.

The CEO's verbal policy may have been intended to foster a family-like atmosphere in the company or, conversely, one that functioned along military-like verbal command lines. Also, the policy may have been the result of more than a modicum of confidence in the personal care taken by each of the company's executives to accomplish the various tasks of his office. A breakdown of the magnitude of the Mount Erebus crash likely could not have been foreseen by the Board or the

CEO. The CEO, in fact, believes that many other corporations operate with the kind of verbal communication practices he encouraged at Air New Zealand. Whether or not that is true, or is true for corporations of a similar size and engaged in similar activities, is a matter to be determined by further investigation.

Importantly, the existence of a policy of verbal communication, etc., as adopted by Air New Zealand, in itself, hardly constitutes a willingness on the part of the Board or the CEO to have such events as this crash occur. It would be grandly unjust to suggest that anyone connected with the administrative structure of Air New Zealand had calculated the trade-offs and opted for the possibility of a disaster such as Mount Erebus rather than a formalizing of the organizational communication system. The fact that the Board and CEO adopted and encouraged the less formal communication system, though having certain of its own problems, is surely an inadequate basis upon which to find them morally responsible for the disaster on Mount Erebus even though causal responsibility focuses on the organizational breakdown.

8. Let us explore a different tack. Strict liability, in the criminal law, is usually applied on the basis of causal identification alone. (Though in some jurisdictions it may be understood in terms of a limited range of conditions of defeasibility, that will not concern us here.) Strict liability is both blind to the reasonableness of beliefs and morally uninterested. Strict liability can, of course, be used to identify the party or parties about which objective liability and moral responsibility questions might be raised, even though they may not be raised in the criminal context. For objective liability we need a standard of reasonableness of beliefs. For moral responsibility we need agency, intentionality. Air New Zealand's organizational and communication structure was causally responsible for the crash and the corporation can be held strictly and possibly objectively liable under the law, for damages. But the matter of the corporation's moral responsibility cannot be resolved by either causal identification alone or by causal responsibility and an appeal to the reasonableness criterion that is embedded in objective liability.

9. The examination of the corporate moral responsibility for the Mount Erebus disaster, however, ought not rest at this level. Commonly, when an untoward event occurs and when the facts will not support an ascription of moral responsibility to the causally responsible party because the event was not intended under the relevant descriptions, the subsequent behavior of the perpetrator is observed to determine whether measures are taken to insure nonrepetition of events of the same kind as the earlier untoward event. If appropriate behavioral changes are not made, a kind of moral reevaluation of the earlier event is made and the perpetrator is held morally responsible for the untoward event regardless of the fact that at the time of its occurrence the perpetrator did not have the morally relevant intention. (Certain excuses, primarily those that claim continuing incapacity or diminished responsibility, defeat the moral reappraisal.)

Suppose, as in the landmark strict liability case of *Regina v. Prince*[3] that a man, named Prince, contrary to law, took a girl who is under sixteen years of age 'away' from her parents and suppose at the time he believed that she was older than sixteen and she gave him every reason to believe so and any reasonable person would have guessed that she was over sixteen. We stipulate that it was no part of Prince's intentions to commit an act under the description for which the law rightly holds him strictly liable. On the basis of the traditional rule of accountability, Prince ought not to be held morally responsible for his illegal assignation. An intuitively appealing, behaviorally orientated, principle of accountability will, however, under certain conditions, license a radical alteration of the finding that Prince ought not to be held morally responsible. I shall call it the Principle of Responsive Adjustment (PRA).

PRA captures the notion that the causally responsible party for an untoward event should adopt specific courses of future action calculated to prevent repetitions. We have strong 'moral expectations', identified by Aristotle, regarding behavioral adjustments that correct character weakness, bad habits, and ways of acting that have previously produced untoward events. But PRA, as I construe it, is more than an expression of such expectations. It allows that when the expected adjustments are not made, and in the absence of supportable exculpating excuses for non-adjustment, the party in question can be held morally responsible *for the earlier event.* PRA does *not,* however, assume that a failure to 'mend one's ways' after being confronted with a harmful outcome of one's actions is strong presumptive evidence that one had intended that earlier event. Under the appropriate description it was not intentional and nothing subsequent to it can make it intentional at the time it happened. PRA incorporates quite another idea. A refusal to adjust one's harm-causing ways of behaving has a second-level effect of associating oneself, morally speaking, with the earlier untoward event. 'Refusal', as used in this context, is an intentional act or acts, and may take a myriad of forms from practised indifference to blatant repetition. The intuition to which PRA must appeal is that a person's past actions (even if unintentional) can be (and often are) taken into the scope of the intentions that motivate that person's present and future actions.

F. H. Bradley wrote, 'In morality the past is real because it is present in I the will'.[4] I construe the PRA as providing an expression of this elusive notion. Certain moral considerations, primarily those that stress the integrity of a lifetime, require adjustments in behavior to rectify flaws of character or habits that have actually caused past evils or, on the positive side, to routinize actions that have led to worthy results. Bradley's idea takes, however, a further reading that exposes the metaphysical foundation of PRA. PRA entails that the intention that motivates a lack of responsive corrective action (or the continuance of offending behavior) affirms, in the sense that it loops back to retrieve, the actions that caused the evil. By the same token, failure to routinize behavior that has been productive of good results divorces

---

[3] *Regina v. Prince*, 13 Cox Criminal Case 138 (1875).
[4] F. H. Bradley, *Ethical Studies* (Oxford University Press, 1876), p. 46.

the previously unintentional action that had good consequences from one's 'moral life'.

Let us try to grasp what this means. Surely, intentions reach forward (like J. L. Austin's miner's lamp),[5] but PRA allows that they also may have a retrograde or retrieval function such that they illuminate past behavior that was unintentional in the relevant moral way. But how can a present intention to do something, or to do it in a certain way, draw a past action into its scope?

Consider Prince's illegal affair with the girl under sixteen. Suppose after serving his punishment, Prince intentionally, and quite deliberately, seeks out other young girls and makes no special attempt to discern their true ages. (Remember he has a penchant for young teenage girls, a penchant, but not an uncontrollable obsession.) In other words, imagine that Prince decides to take himself down to the local high school and pursue another teenage girt to whom he has taken a fancy. Prince's intention with regard to this continuation of his behavior with such girls is formed within a personal history that includes his conviction in *Regina v. Prince*. Prince is aware of his past, indeed, in a Lockean sense, his past behavior and its subsequent punishment is, we should expect, a part of his current consciousness. The memory of those events is coconscious within his mental history and that mental history constitutes, in conjunction with his concerns for his well-being in the future or his life plan, his identity as a person, at least in a relevant moral and legal sense. If, subsequent to the commission of the crime, Prince had taken precautions to insure that he learn the ages of the girls he courts, etc., we would allow his ignorance of the age of the girl in the legal case as an exculpating excuse for moral purposes (although he must bear the punishment for the strict liability offense). But, in our extension of the story, Prince makes no responsive adjustments in his behavior that would have the effect of not putting him in the position of repeating the offense. In fact, he embarks upon a course that could very well lead to another violation of that same law, though whether or not it does is immaterial to PRA. By PRA, we are permitted to claim that Prince's subsequently manifested intention to continue his romantic pursuits of underage girls constitutes an affirmation of the strict liability violation behavior. In other words, by virtue of a retrieval function in the subsequent intention, Prince may be legitimately held morally responsible for the strict liability offense.

A second example may help to fortify the intuition. Suppose that Sebastian gets drunk and drives his car onto Quincy's property. Let us assume that Sebastian had no intention of damaging Quincy's property, or to get drunk. After regaining sobriety and learning of his misadventures, Sebastian, who is not yet an alcoholic, subsequently and quite deliberately returns to the local pub and proceeds to get himself roaring drunk. Again we should say that Sebastian's past is known to him, at least he is aware of the fact that he got drunk on a certain occasion and destroyed Quincy's property. Yet, he embarks upon a similar course again. Though we would have excused him from moral responsibility for the damage he did on the first spree had he subsequently modified his behavior, he did not do so and, by PRA, he is making

---

[5] J. L. Austin, *Philosophical Papers* (Oxford University Press, 1961, 1970), p. 284.

the crucial past behavior, not an out-of-character happenstance, but very much in character and hence something for which, as Aristotle would say, he may be held morally accountable. Seen in this light, PRA captures (at least to some extent) the Aristotelian idea that we do not blame people for unintentionally "slightly deviating from the course of goodness",[6] as long as they do not subsequently practice behavior that makes such deviations a matter of character.

PRA insists that moral persons learn from their mistakes. "It was inadvertent or a mistake" will exculpate only if corrective measures are taken to insure nonrepetition.

The most radical element of this analysis surely is that which provides for a retrieval of past unintentional behavior in a present intention to do something. Although this strikes me as quite consistent with common intuitions, a more technical account is surely wanted. That account, however, is easily at hand. Intention, as we know, is an intensional causal notion. As such, it is referentially opaque so that the aspect of the event described makes all the difference with regard to intentionality. We intend under event or action descriptions and we redescribe actions and events as licensed by certain rules. To say that someone intended to do something is to say that there is at least one proper description of some event under which that person acts. Act descriptions have a well-known feature that Joel Feinberg once called the 'accordion effect'. Like the musical instrument, the description of a simple event can be expanded in different directions to include causal and other related aspects that might themselves be treated as separate events for different purposes. For example, the act of pulling the trigger of a rifle might, through a series of redescriptions, be expanded to the description 'the killing of the judge'. Accordions, of course, can be drawn apart in both directions. The description of Sebastian's present act of getting drunk may be associated to his past action by the ordinary relations 'like yesterday', 'as before' or 'again'. It is the case that Sebastian, in our story, intends to get drunk again. The action intended under that description dearly retrieves the previous behavior, though it certainly does not make the previous behavior intentional at the time it occurred.

PRA, however, demands a bit more than this because Sebastian could offer the plea that he had not intended to get drunk under such an 'accordioned' description at all. He only intended to get drunk simpliciter. We may, however, reject this plea on the grounds that unless he is of diminished mental capacity or suffering from amnesia, etc., his grasp of what he is doing is made within a mental or personal history that is not present-specific. The descriptions of events under which he intends his actions are formed within that history. 'I intended only to get drunk, not to get drunk again', is, in this context, unintelligible. There are limits on excludability by appeal to semantic opacity in intension. If Sebastian, quite intentionally makes his way to the pub in order to get roaring drunk, he intentionally goes to get drunk again, or like yesterday, etc., and his doing so affirms the previous episode insofar

---

[6] Aristotle, *Nicomachean Ethics* (Indianapolis: Bobbs-Merrill Company, 1962), p. 51.

as it takes it into the description captured in the scope of the intention. That is what PRA requires.

The application of PRA to cases like our imagined extension of *Regina v. Prince* and drunken Sebastian can be generalized to the position that although a person may not have access to the relevant information at the time of an action that produces a bad or harmful outcome and so could not reasonably be said to have intended that outcome, that person may still be held morally responsible for that outcome if he or she subsequently intentionally acts in ways that can reasonably be said to be likely to cause repetitions of the untoward outcome.

No strict set of temporal closures need be applied to PRA. There is no statute of limitations. For example, 'moral enlightenment', many decades after an event, may demand reevaluation of an action or an outcome that was not originally thought to be bad or harmful and PRA will require moral accountability of the perpetrator if, after enlightenment, no behavioral adjustments are made.[7]

PRA has another important intuitively appealing aspect. Suppose that we think of all of those acts for which a person can be morally blamed or morally credited as exhausting that person's moral life, the biography that can be morally judged or evaluated against a standard of worth or virtue, as Aristotle would have it, "in a complete life". PRA may incorporate originally nonintentional pieces of behavior into a person's moral life because PRA does not let persons desert their pasts. It forces persons to think of their moral lives as both retrospective and contemporaneous, as cumulative. Moral persons cannot completely escape responsibility for their accidents, inadvertent acts, unintended executive failures, failures to fully appreciate situations, bad habits, etc., simply by proffering standard excuses. PRA, in fact, defines the boundaries of acceptability of pleas of the form "it was unintentional". If I am right, the ordinary notion of moral responsibility is far grander than usually described. It operates over more than isolated intentional acts considered seriatim. The moral integrity of a person's life depends upon a moral consistency that is nurtured by PRA.

10. Let us now return to the tragic crash of TE- 901. The Royal Commission reports, as already mentioned, that when Air New Zealand became aware of the crash, it proclaimed the pilot error theory. Its CEO also ordered that only one file be compiled, to be kept in his possession, of all of the airline's pertinent information regarding TE-901, and all other documents regarding the flight (duplicates, etc.) be shredded. [#338 and 373] This may appear a suspicious move, but it is not indefensible, for the airline wished to avoid trial by the press that could affect future settlements with relatives of the victims. In fact, PRA does not direct immediate attention to those and other seemingly irregular activities of the airline management immediately after the crash. The Royal Commission, I think, is most unkind in its description of the post-accident behavior of the senior executives of Air New Zealand. It refers to the testimony of airline exec-

---

[7] This was pointed out to me by Professor Lisa Newton.

utives as "an orchestrated litany of lies". [#377] (That finding has, however, been recently overturned by the New Zealand Court of Appeal.)

The crucial question is, "Did Air New Zealand, when fully apprised of the extremely strong case made in the Commission Report, move to make adjustments in its internal administrative systems?" Has Air New Zealand restructured its standard operating procedures, especially in the Flight Operations Division, to correct the deficiencies outlined in the Report? The answer is "No".

Rather than redesign its policies to incorporate a way to ensure that information within and among the sections of the Flight Operations Division is placed in the proper hands, Air New Zealand has continued in court, interviews, and company documents to defend its old procedures, structure, and verbal communication policy. Rather than accept the findings of the Commission Report, Air New Zealand continues to insist that the primary cause of the disaster was pilot error.[8] They have, it should be noted, discontinued all flights to the Antarctic, but it is not solely to such nonregular flights that attention should be drawn. Flight Operations still functions just as it did before November 28, 1979 for all scheduled flights.

PRA is brought into the analysis of the moral responsibility for the crash if we (1) substantially accept the findings of the Commission, and, (2) confirm that Air New Zealand has taken no responsive adjustment measures to correct the organizational flaws identified by the Report as "the single effective cause of the crash" [#399], regardless of whether they have not had any more recent serious crashes. On PRA and given a fair reading of the facts, it seems clear that Air New Zealand should be held morally responsible for the Mount Erebus crash. Its post-crash and post-Report behavior manifested the intention to continue its crucially prone-to-defect structural and procedural policies, and that intention retrieves within its scope, the corporate actions (or the actions of corporate personnel in its Flight Operations Division) that "programmed the aircraft to fly directly at Mount Erebus and omitted to tell the aircrew". [#494] Again the Commission Report is quite explicit. "That mistake is directly attributable, not so much to the persons who made it, but to the incompetent administrative airline procedures which made the mistake possible." [#393] Air New Zealand's failure to adjust, on PRA, provides the basis for the justification of an ascription of moral responsibility for the tragic crash to the airline.

11. The Mount Erebus case may be paradigmatic of a large class of cases in which questions of corporate moral responsibility arise, not because it is dramatic and involves great loss of life, but because the focal event is not originally corporate intentional. Although I think that PRA is a basic principle of morality and hence is applicable to all persons, it seems to be particularly appropriate in morally evaluating corporate actions. Corporations, through their standard operating procedures, may actually have a greater capacity for reactive adaptation than do

---

[8] This remains true at the time of this writing [1984], as reported by John Braithwaite and W. Brent Fisse who have recently interviewed Air New Zealand executives while researching a forthcoming book on the effects of publicity on corporate policy.

human beings. (But that is only a matter of degree and, I think, not a matter of any consequence.) Often, corporate personnel or subcorporate units at all levels of an organization do things that result in untoward events, and a higher subcorporate unit, or the corporation itself, must decide on a responsive course of action. PRA allows the incorporation of the actions of personnel in the intentions of the corporate body and hence a finding of corporate moral responsibility.[9] But by the same token, the appropriate corporate internal adjustment in response to the revelation of the unintentional (at least at the corporate or subcorporate unit level) causing of an untoward event preserves corporate moral blamelessness in the event. PRA gives us an explication of the corporate excuse that it was just a foul-up in the organization, or the selfish dealings of an individual, or a misdeed by someone, who, though he may have believed he was acting for the corporation was actually pursuing his idiosyncratic conception of the corporate interests and corporate policies.

Although I have only scratched the surface here, I hope I have provided some good reasons to support the view that the Principle of Responsive Adjustment, Aristotelian in origin, is most comfortably at home in our ordinary idea of moral accountability and should be given more serious consideration by philosophers generally interested in ethics and specifically concerned about the notion of corporate responsibility.

---

[9] Organizational defects are one of the most common causes of corporately caused untoward events and they are also responsible for many criminal violations (particularly in the advertising area). A case in point is reported in Andrew Hopkins' study for the Australian Institute of Criminology on *The Impact of Prosecutions Under the Trade Practices Act* (April 1978). The Sharp Corporation of Australia falsely advertised that its microwave ovens were approved by the Standards Association of Australia (SAA). The ovens had been approved by the New South Wales Electricity Authority according to standards set for electrical equipment by the SAA. But no SAA standards existed at the time for microwave ovens. The Sharp sales manager, who authorized advertisements, claimed that although his technical staff would have been aware of the distinctions in the standards and authorities, that staff is not, in the standard corporate procedure, consulted on such advertising and he was ignorant of the relevant distinctions. 'Sharp's failure was a failure to involve the appropriate technically qualified people in checking the advertising copy before publication... the offence was not intentional' (pp. 5–6). In response to prosecution, Sharp changed its advertising procedures to ensure that technical staff check copy before it is published to prevent any recurrence of the false advertising. Sharp's responsive adjustment is sufficient to rule out a finding of moral responsibility in the matter, whereas a failure to alter the corporate procedures, given PRA, would have provided a clear ground on which someone might claim, as one judge in the case did, that the advertisement was a 'gross and wicked attempt to swindle the public', an intentional corporate action meriting moral reprobation and animadversion.

Chapter 4
Science as a Profession: And Its Responsibility

Harald A. Mieg

Abstract Scientific responsibility has changed with the successful professionalization of science. Today, science is a privileged profession, one with a (tacit) management mandate for systematic knowledge acquisition. Within this framework, science acts with responsibility. This chapter reflects the responsibility of science in the German context. After Wold War 2, the extraordinary responsibility of scientists, which C.F. von Weizsäcker emphasized, referred to a specific phase in the institutional development of science, termed *scientism* ("science justifies society," science as religion), and corresponded to an elite responsibility. Today, one responsibility of science as a profession is to safeguard and develop scientific standards. This also concerns, on the one hand, the self-organization and control of science as a profession and, on the other hand, the communication of science to society. As a professional scientist, one has two responsibilities, the commitments to good science (professional ethics plus co-responsibility for the development of science as a profession) and civic responsibility. Due to their special knowledge, the civic responsibility of the scientist differs from that of other professionals. This chapter introduces science as a profession and presents an integrative notion of responsibility, also shedding light on the social responsibility of science.

Introduction: Knowledge Is Power?

In April 1957, a group of 18 scientists published the Göttingen Declaration, which expressly opposed plans for nuclear armament in the Federal Republic of Germany. The declaration cited the enormous risks associated with nuclear weapons and "the responsibility for the possible consequences" that scientists must bear in this context. Among the signatories were several Nobel Laureates such as Otto Hahn and Werner Heisenberg. The declaration had a lasting effect by leading the public debate. Göttingen was intentionally chosen in reference to the protest of the

H. A. Mieg (✉)
Institute of Geography, Humboldt-Universität zu Berlin, Berlin, Germany
e-mail: harald.mieg@hu-berlin.de

© The Author(s) 2022
H. A. Mieg (ed.), *The Responsibility of Science*, Studies in History and
Philosophy of Science 57, https://doi.org/10.1007/978-3-030-91597-1_4

Göttingen Seven from 1837. Seven Göttingen professors, including the Brothers Grimm, had protested against the abolition of the liberal state constitution. The king dismissed the professors, but the protest had set a lasting sign for political development.[1]

A parallel could be seen today in the discussion on climate change. Here, too, scientific research has fired up a public debate, even though the culpability of science in the development of the problem is limited and consists of the general interaction of science with industrialization. A major distinction from the discussion on nuclear weapons is that the discussion on the scientific side cannot be tied to specific names. Rather, it is an institution close to science—the IPCC (Intergovernmental Panel on Climate Change)[2]—that has driven the translation of science into politics and alerted the global public. This is an example of scientific responsibility as corporate responsibility. The IPCC forms global teams of researchers in compiling its reports and attempts to achieve consensus within the scientific community.

The thesis presented herein is that: The question of responsibility has changed with the successful professionalization of science; science has acquired self-control as a profession. The professionalization was sustained by an enormous expansion of science, as can be seen from the number of universities and professorships. While in 1950 there were approximately 5500 full-time professorships in West Germany, by 1995 there were already about 34,000 (an increase of more than 500%). For the reunified Germany this number rose again from 37,672 in 1995 to 47,568 in 2017 (an increase of about 26%).[3] With professionalization, many (but not all) ethical questions in science are brought up for discussion and regulation within the profession and translated into guidelines for good practice. This is more successful in central sectors of the scientific community, such as universities, than in industrial laboratories. It also succeeds better in subject areas whose progress depends on scientific and technical methods (for example in human genetics) than in open fields such as IT development, which are less dependent on professionalized science.

The phrase "Knowledge is power" is commonly attributed to Francis Bacon (1561–1626). Because of the power that knowledge represents, science should assume corresponding responsibility. Knowledge as power has always applied to all areas of life, from warfare and corporate governance to educational issues.

---

[1] In the context of 1837, the Göttingen Seven showed civil courage with their protest ("role-discrepant responsibility", cf. Mieg, 1994b, 2015). The reaction of the University of Göttingen was one of anticipatory obedience ("role-conformant responsibility"): it distanced itself from the Göttingen Seven; it was concerned about its status and the welfare of its professors and students. In contrast, in 1957, this seemed neither opportune nor necessary for the university.

[2] The Intergovernmental Panel on Climate Change (IPCC) is an advisory body of the UN Environment Programme (UNEP) and the World Meteorological Organization (WMO), and was established in 1988 with its headquarters in Geneva. Between 1990 and 2021, the IPCC has published six globally leading Assessment Reports on climate change, encompassing all forms of global climate research; separate peer review procedures have been developed for this purpose.

[3] Comparison for West Germany according to DFG (2013, p. 42). Comparison for Germany as a whole according to BMBF (2018); corresponding figures for scientific personnel: 152,401 in 1995 and 249,535 in 2017.

Knowledge is not a privilege of science. Science, on the other hand, has professionalized (and largely monopolized) the systematic acquisition of knowledge. The professionalization of science means a certain relief for the individual for the sake of specialization. Today, one does not need to be a hero in order to do science; it is enough to have qualifications. Then the old legal principle "ultra posse nemo obligatur" applies: no person can be obliged to do more than they are able to do. The relief of the individual is accompanied by a new responsibility of science as a profession. This will be examined in more detail in the following sections.

Science as a Profession

Today, science is a privileged profession and first and foremost responsible within this framework. The privilege of science consists of its *autonomy* as a profession, i.e., a right to self-organization.[4] Science shares this privilege with other professions such as medicine and architecture. In Germany, one expression of the self-organization of science is the DFG, the German Research Foundation.[5] The DFG receives state funding with which it supports research projects, and organizes itself in line with the university subjects. A general autonomy—which would amount to a certain degree of self-sufficiency—is not close. For science, at least as far as the universities as the central expert organizations are concerned, is dependent on state support.

A prerequisite for autonomy is the authority to evaluate performance. What qualifies as good science is defined by science itself. Science has a monopoly over its own *performance evaluation* and quality definition. For this purpose, the peer review process has become established in the scientific community. A scientific contribution—be it an article or a project—is evaluated by at least two experts in the field. Science as a profession is almost as old as architecture. In a way, science has always been there. Aristotle systematized science and can be considered the first

---

[4] I have presented my understanding of professions and professionalization in various publications (e.g., Mieg, 2005, 2018). A profession is a profession privileged by autonomy. From the perspective of the sociology of professions, today's professions are dependent on the use of abstract, scientific knowledge (cf. Mieg & Evetts, 2018). Moreso: professions secure for themselves the responsibility for abstractly defined problem areas, e.g., health, architecture, law... (see Abbott, 1988; Freidson, 2001). The abstractness of the knowledge base and the derivation of measures (for example in medicine) makes access for other professions more difficult, thus ensuring autonomy. A profession "monopolizes" the definition of problems and solutions in its field (What is Alzheimer's Disease? What are suitable therapies for it?). This is all the more important as professions have given up many functions or privileges over the course of time or else have never really achieved them, e.g., the control of the market or the training of their own young people.

[5] The DFG (German Research Foundation) is a registered association and was founded in 1920 as the "Notgemeinschaft der deutschen Wissenschaft" (Emergency Association of German Science) (at the suggestion of Fritz Haber), renamed DFG in 1951. The DFG provides financial support for research projects, amounting in 2018 to 3.4 billion euros (Annual Report 2018) and also advises policymakers.

scientist. However, a real version as a profession—a social closure—occurred only much later, in the nineteenth century for architecture, in the twentieth century for science.

Science is a special profession,[6] if only because one can still rightly ask whether it is a unified profession at all, because science breaks down into *disciplines*. Moreover, through the universities, it carries out subcontracting work for other professions, such as doctors or architects. Rudolf Stichweh depicted the productive relationship between discipline and profession[7]: Medicine exists as a scientific discipline; its task is to develop the scientific foundations and provide appropriate training. On the other hand, there is the medical profession; it comprises the practicing medical profession, whether in individual practices or clinics. This division of tasks between discipline and profession is obvious and would also fit lawyers, with the law as the discipline, and with advocacy and judges as their applied, professional side. However, what about other subjects, such as physics or philosophy? In those fields, independently practicing philosophers or physicists are the exception.

A strong indicator of professionalization is the *formation of associations*.[8] In subjects that have both a strong scientific and practical side, there are usually two associations, one for research and one for practice. In psychology, for example, we find the DGPs (German Psychological Society, founded in 1904) as a scientific society and the BDP (German Association of Psychologists, founded in 1946) as a professional association. As mentioned above, there is a lobbying body for science as a whole, the DFG, which absorbs changes in subjects and disciplines.[9] The professional exchange relevant for research is carried out through the scientific societies (e.g., DGPs for psychology). In contrast to the Hartmannbund, i.e., the German Professional Association of Doctors, or the BDP, i.e., the German professional association of psychologists, the DFG is not a professional association in the classic sense, and one cannot become a member as an individual researcher. In addition to

---

[6] Only a few sociologists of the profession discuss science as a profession. These include Ben-David (1972), who discusses the historical development of the role of scientists, and Oevermann (1996, 2005), who sees the production of truth (and treating epistemic crises) as an essential service that needs to be "professionalized."

[7] Stichweh (1994).

[8] Freidson (1986).

[9] I am aware that my argument for the late professionalization of science (still) stands on weak legs. Looking at the professional associations—scientific societies—science as a profession was by no means late. The Society of German Natural Scientists (today the Society of German Natural Scientists and Physicians, GDNÄ) was founded in 1822, the British Association for the Advancement of Science (today the British Science Association, BSA) in 1831, and the American Association for the Advancement of Sciences (AAAS, today the largest of its kind worldwide) in 1848. Only the AAAS is still a professional association for scientists. The GDNÄ and BSA today serve the purpose of promoting scientific understanding in society. The national academies are much older again, including the Leopoldina (founded in 1652 as Academia Naturae Curiosorum), the Royal Society (1662), and the Académie des sciences (1666). These scholarly societies serve the exchange of expertise as well as governmental advice and are generally elitist, i.e., they are not open to all scholars.

the transfer of research funds, the DFG limits itself to one core task of professional self-organization: the definition of quality standards for science.

The final professionalization of science only took place in the last decades of the twentieth century. This delay was due less to organizational weakness than to its social, even *ideological* strength.[10] In the twentieth century, science often served as a potential state religion or at least as a vehicle for justifying politics, in the USA as well as under National Socialism in Germany and Soviet Communism. Science was elitist. The turnaround was triggered by the crises that began after 1970. The oil crisis of 1973 made it clear that science-based, determinant planning was not possible. The world turned out to be more complex, more unpredictable than could be grasped through science. Nevertheless, scientific methods had proved indispensable for industrial innovation. In the case of engineering, chemistry, and medicine, this had long been known. What was new was that some, inevitably, applied science to fields far removed from the natural sciences (such as market research) that produced practical success. After the Second World War, science of all kinds was increasingly in demand by international institutions such as the OECD and WHO. Research on national innovation systems, which was funded under the aegis of the OECD, made it clear that national growth—measured by GDP—is linked to investment in research and development. The lever was now no longer in the direct application, the transfer, of science to politics and business, but in the promotion of the scientific system as a whole. The resurgence of globalization and digitalization since 1990 has finally put science on the path to professionalization. Science has always been global but, with the Internet, science in particular achieved a completely new acceleration and expansion potential.

Professionalization is expressed in *social closure*. A profession internalizes the discussion and processing competition for problems of a certain type. Medicine offers a vivid example. Many tasks that are now taken for granted as part of medicine previously lay in other areas of responsibility, for example, the internment and treatment of the mentally ill (psychiatry) or dental treatment. Even the treatment of external wounds was once not part of a doctor's role. Over the years, more and more tasks have been successfully internalized, as has the professional competition. If the competition once took place between medicine and other professions, such as barbers (once responsible for wound and dental treatment), this is now shifting to medicine itself as internal competition between medical specialists.

Such closure processes are never finished; the treatment of moral questions shows this. Each profession develops a *code of professional conduct*. The Hippocratic Oath of the medical profession is quite old. Corresponding ethical codes for science were not established until long after the Second World War. The example of

---

[10] The idea of science, or science as a system, is historically very successful. Even if science does not monopolize truth and knowledge, its promise of truth production is attractive to people of all times. This (in my opinion) explains the elevator effect in higher education: No sooner had a new form of higher education 'latched onto' the scientific system (i.e., as soon as it followed scientific logic), than it sought recognition as a university. This was true for the technical universities of the nineteenth century and applies to the universities of applied sciences of the twenty-first century.

psychology shows how dynamic are professional codes of ethics. The first code of the American Psychology Association (APA) was created in 1953.[11] A revision seemed necessary in 1973, which was preceded by 7 years of discussion,[12] not least in response to the Milgram experiments (concerning obedience to authority figures, in which participants incorrectly believed that they were subjecting test subjects to electric shocks). A further revision took place a few years ago, when it became clear that APA psychologists were being employed to refine torture methods at the Guantanamo Bay detention camp. It is important to note the role of internalization: ethical questions confronting science as a whole are converted into rules of conduct for individual scientists.

Responsibility and Science in Former Times: From (Paid) Hobby to Scientism

People have always conducted research. The observation of nature and meteorological phenomena has probably always been important, at least since the beginning of agriculture and human settlement. Science as a systematic of research and its knowledge has existed in our civilization since the times of Greek antiquity. But if scientists have existed almost as long as doctors and architects, science has long been a *non-plannable* profession, unlike craftsman, mercenary, or nun. One became a scientist through a self-chosen secondary occupation, often in the service of the church or in the leisure time available to the nobility (the gentleman scientist). Notable examples were the busy judge Pierre de Fermat (1607–1665), who in his free time formulated groundbreaking propositions and riddles of mathematics; the farmer Johann Georg Palitzsch (1723–1788), who undertook astronomical research and was the first to observe the return of Halley's Comet; and the nobleman Sir Henry Cavendish (1731–1810), who secretly conducted physical experiments and, for example, discovered the element hydrogen. In the late Middle Ages and early modern times, scientists were also employed at royal courts, even for astrological advice. Science for purposes of war was always in demand. Archimedes (287–212 BC) not only described the laws of leverage, but also invented catapults to repel enemy ships. Leonardo da Vinci (1452–1519), the great universal scholar, sought support and employment with the promise of inventing new types of weapons.

For a long time, science was left to self-selection. Only a few had the talent and opportunity to devote themselves to science. This only changed with the beginning of modern times. Francis Bacon (1561–1626) exemplifies this new beginning. He vehemently advocated the experiment as a means of science. This was new insofar as science in the sense of Aristotle was understood as the observation of nature, paradigmatically implemented in astronomy. Experimenting, on the other hand,

---

[11] Cf. Joyce and Rankin (2010).
[12] Cf. Stark (2010).

meant controlled change. Not only that: according to Bacon, science should serve *progress*. This gave science its own responsibility.

The twinning of science and progress determined the development of science until the twentieth century. Industrialization was accompanied by an upswing in science. The rise of the German chemical industry is symbolic of this, going hand in hand with the upswing in chemical science and producing global corporations such as Bayer and BASF. The Nobel Prize[13] in Chemistry, which has been awarded since 1901, went to a German researcher every second time during its early years. Scientific research was institutionally anchored in the newly established *Kaiser-Wilhelm-Gesellschaft*, founded in 1911, today known as the *Max-Planck-Gesellschaft* (Max Planck Society). Science gained a new, prominent position and thus also responsibility—the responsibility of an *elite*. For science was still not really understood as a profession, but rather as the pursuit of excellence. The success of the *Kaiser-Wilhelm-Gesellschaft* was also based on this. Harnack, its first president, had introduced a principle named after him: An institute of the Kaiser Wilhelm Society was not conceived as an institution with specialist research tasks; it was not about a task, but about a person. The principle was that a researcher received generous funds to be able to build an institute for their research. The idea of the social positioning of science still concerned not the profession but rather the vocation.

While the German term for a scientist, "Wissenschaftler," was still based on a nineteenth-century feuilletonistic joke title, science took an elitist turn in the twentieth century. Science became a substitute religion, *scientism*, and took on a state-defining function. The National Socialists (with the support of the DFG) were able to rely on science, as was the Soviet Union, and the USA long equated democracy with science.[14] And everywhere this means progress at the same time. C.F. von Weizsäcker's appeal to the responsibility of science belongs in this context. It is an elite responsibility, shaped by expectations of technology and planning. Science explores spaces of possibility. According to C.F. von Weizsäcker, the scientific elite had to limit these spaces. However, scientism also always means the possibility of national appropriation, right up to the idea of a specifically National Socialist or Soviet 'science.'

The inner professional principle of science has always contradicted national appropriation, namely to exchange information and thus create *transparency*. National borders do not play a role here. The requirement to publish corresponds to an inner necessity of science: only what has been made known to colleagues counts.

---

[13] The Nobel Prize has assumed an important evaluative function for science. Since it is only awarded in a few disciplines, equivalent prizes were created, e.g., the Fields Medal in mathematics and the Pritzker Prize in architecture. Control of evaluation is central for a profession (Mieg, 2018).

[14] This is clearly expressed in the US approach to critical thinking (cf. Glaser, 1941; Ennis, 2011). Critical thinking is largely modeled after scientific thinking: the critical thinker "[j]udges well the credibility of sources,... [f]ormulates plausible hypotheses, ...[p]lans and conducts experiments well" (Ennis, 2011, p. 12). According to Ennis (2011), "the survival of a democratic way of life depended on the critical thinking of the voters" (p. 5).

The discovery of nuclear fission became a globally conscious possibility for the construction of atomic weapons through its scientifically inevitable publication by Otto Hahn and Fritz Strassmann in January 1939. We can't really stop science, but we can guide it and create framework conditions. This applies both to nuclear research and to modern medicine.

The fact that science is now professionalized has to do with the work of two other players, the universities and industry. Even though many famous scientists taught at universities, the symbiosis of *university* and science, as we know it today, is not yet old. Historically, universities are autonomous educational institutions. The university defines itself through the community of teachers and learners, professors, and students. For centuries, research has taken place outside universities, e.g., in academies or privately. The modern research university is an invention of the early nineteenth century that was successfully exported from Germany to the USA. The expansion of the university sector—academization—has created new opportunities. Both profit from the connection between university and science. For science, the universities offer employment with sufficient professional freedom; in turn, the universities gain legitimacy and reputation through science.

No less important is the role of *industry*, where there is a constant demand for cutting-edge science as a means of competitive advantage. In Germany, the bulk of investment in research and development comes from industry.[15] The uptake of science by university and industry naturally has its price: in academic it is the obligation to teach; in industry it is marketability. Industry has become the hoard of scientific technology visions that were once nurtured by governments. This includes the great pioneering visions, for example in space travel, as well as the mechanization of entire living environments, for example as a *smart city*. Here, industry benefits from the success of scientifically supported market and advertising research. For industry, science itself is a marketable vision. In the slipstream of this new demand and challenges from universities and industry, science has become more professional.

Two Individual Responsibilities of Science: Professional Ethos and Civic Responsibility

In 1957, shortly after the Göttingen Declaration, C.F. von Weizsäcker gave a lecture for student bodies on the "Responsibility of Science in the Atomic Age." He saw the responsibility of science within the framework of the intertwining of science and technology, which he discussed under the title "Plan und Mensch" ("Plan and Man"): "If it is romantic to want to throw off technology, it is conversely childish to

---

[15] In Germany, more than two-thirds of research and development expenditure is financed by industry (see BMBF, 2018).

do everything that is technically possible."[16] He demanded in particular that we do not dismiss the opportunity for reflection. Our responsibility in the technical world therefore means at least: we must learn to remain human in the midst of planning and apparatuses. Or: "Correct, responsible planning and technology have distance from the apparatus."[17] In this way responsibility—according to C.F. von Weizsäcker—can become concrete. With regard to the Göttingen Declaration, he explained: "We had to turn to where we have a direct civic responsibility, namely to our own country…"[18] The responsibility to which C.F. von Weizsäcker appeals here is civic, and the contact is with the state. The obligation of science associated with this is supererogatory, i.e., it goes beyond what is normally expected.[19] The exemplary responsibility achieved by the Göttingen Declaration is ultimately the *responsibility of the elite*.

For scientists today, there are two kinds of responsibility (see Fig. 4.1). On the one hand, the commitment to the *ethos*, which is based on the scientific,

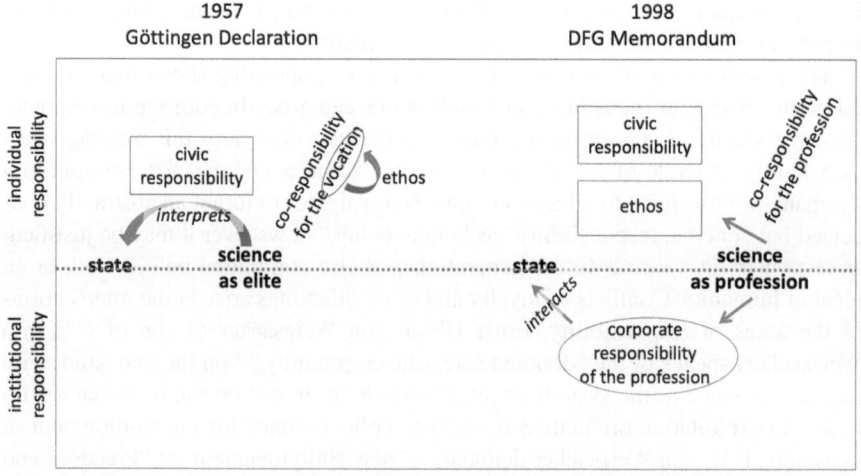

Fig. 4.1 Change in responsibility of science: from the responsibility of an elite (whose members have to interpret their civic responsibility) to professional responsibility. The (vocational) ethos is formalized in the course of professionalization and is no longer left solely to the professional understanding of an elitist (sub-)group. The frame of reference here is the sphere of responsibility of the nation state. Universal, moral, family, or corporate responsibility is not represented here. What is relevant, however, is co-responsibility for the vocational group or profession; in the course of professionalization, this is also formalized, e.g., within the code of ethics

---

[16] Von Weizsäcker, C. F. (von Weizsäcker, 2011), p. 10 [translated].

[17] Op. cit. p. 13 [translated].

[18] Op. cit., p. 26 [translated].

[19] Supererogatory means: beyond the expected level—overly excessive; morally valuable, because self-sacrificing, but therefore not obligatory. One is not obligated to put oneself in danger to rescue another from a burning house (unless one's profession is a firefighter).

professional role and is therefore coupled with the co-responsibility for science as a profession (not least in order to further develop the ethos). On the other hand, there is *civic responsibility*, to which C.F. von Weizsäcker also appealed, and which varies depending on the type and extent of knowledge, in accordance with the understanding of knowledge as power. In his lecture, C.F. von Weizsäcker explicitly addressed the scientific ethos only once. He commented on the publication by Hahn and Strassmann of their findings on nuclear fission: "Publication is considered a duty in science; it means subjecting one's views to the control of colleagues."[20] He did not even suggest a possible conflict with the interests of the Nazi war preparations, which would have suggested keeping secret the progress in nuclear physics.[21] When C.F. von Weizsäcker states that the personal responsibility of the natural scientist "corresponds to the practical importance of his subject,"[22] then the principle of "knowledge is power" is reflected here. The result is again civic responsibility. In today's highly differentiated societies of the twenty-first century based on the division of labor, however, the general civic responsibility has been reduced almost to the unconditional obligation to pay taxes, which relieves us of further obligations. Even participation in the political process is not a 'must' in the context of civic responsibility, but a 'should' and counts as voluntary.

The juxtaposition of scientific ethos and civic responsibility shows that there are different *areas of responsibility* in which duties can arise. In addition to responsibilities to science as a profession and those to the state or society, this includes one's own family or circle of friends as well as the business context, for example in a company or university, to which one must be loyal. Not to forget an abstractly conceived but concrete responsibility "as human being," howsoever it may be justified: as responsibility towards future generations, or else committed before God, or an ideal of humanity. Conflicts of loyalty and other dilemmas arise at the intersections of the areas of responsibility. Ernst Ulrich von Weizsäcker (a son of C.F. von Weizsäcker) speaks of the "doubled scientific community,"[23] on the one hand, with regard to science as the system oriented towards truth, and on the other hand with regard to exploitation orientation in order to collect money for the employment of assistants. E.U. von Weizsäcker demands a "new Enlightenment"[24]: "Freedom and democracy must be protected *against* the market."[25] What is needed is a "new balance between state and economy."[26] Because, as mentioned, atomic bombs and embryo research, including the visions of technology, have become an industrial pursuit. The good—or difficult—thing about this situation is that science, too, has

---

[20] Op. cit. p. 17 [translated].

[21] For Otto Hahn as a person, such a conflict would have been unthinkable anyway. In 1934, he resigned from the teaching staff of the University of Berlin because of the exclusion of Jewish colleagues, and in 1938 he helped Lise Meitner to escape Germany.

[22] Op. cit. p. 15.

[23] Von Weizsäcker, E. U. (von Weizsäcker, 1986), p. 221 [translated].

[24] Op. cit. p. 228 [translated].

[25] Op. cit. [translated].

[26] Op. cit. [translated].

emancipated itself from the state and is a profession that represents its own interests alongside the state and the economy. No wonder that in this mixture of different loyalties, the role of the *whistle blower* receives new attention. With the legal protection of *whistle blowers* an element of civic responsibility is brought to bear. Whether a "new Enlightenment" can lead the way out of such dilemmas remains questionable.

The Scientific Ethos: From Merton to the DFG

The American sociologist Robert K. Merton was one of the first to deal explicitly with the ethos of science. In his 1942 essay "The normative structure of science," he identified four principles or "institutional imperatives" that made up the scientific ethos. Firstly, universalism: findings apply regardless of who makes them; they only have to prove themselves before the existing system of findings. Second, "communism" (original quotation marks): Scientific knowledge belongs to everyone and becomes a public good. Third, disinterestedness: By this Merton means that science has no primary interest in exploitation, but aims for knowledge in itself. Merton notes that science differs in this respect from the other classical professions such as medicine and law, since science lacks the usual client relationship. Fourth, organized skepticism, i.e., a norm for restraint in judgment until empirical evidence is available, and for questioning assertions.

Do these principles still apply today? Helmut F. Spinner spoke in connection with professional ethos, as defined by Merton, as "qualified superethics for privileged special environments."[27] Increasing professionalization is accompanied by increased pressure from employers, i.e., universities and non-university research institutions. It is precisely the characteristic of disinterestedness that comes under pressure; doing science and remaining in the science system become existentially relevant. It is not only a matter of reputation but also of being able to support a family as a scientist—just as in other normal professions. The expansion of the science system makes it necessary to simplify and formalize performance assessment. Because the question: "What is good knowledge?" can often only be answered long after a research project has been completed. It is therefore hardly surprising that Schurr, for his considerations on an *academic code of ethics*, placed the measurability of standards in the foreground. Schurr's list of criteria begins with *auditability* and ends with *enforcement*.[28] The price that science paid for professionalization in order to achieve freedom and responsibility—like other professions—was the quantitative evaluation of performance through the number of publications and citations, etc.

---

[27] Spinner (1985), S. 56 [translated].

[28] See Schurr (1982). Schurr relies on non-scientific control bodies (p. 332), e.g., the press. After all, science was acting in the public interest. Schurr (1982) considers the question of scientific ethos from the perspective of educational science.

With the normalization of science as a profession, it became inevitable that here and there fraud came into play or became notorious. Even for Merton, in 1942, this seemed irrelevant. Merton justified disinterestedness as a principle of the ethos in science by hinting to the "virtual absence of fraud."[29] In 1992, however, the *National Academy of Sciences* (NAS) of the USA took a step that other professions had already taken, that is to establish a code of practice.[30] The DFG followed in 1998.[31] The NAS Code of Good Scientific Practice was aimed at individual researchers. Recommendation 1 was:

> Individual scientists in cooperation with officials of research institutions should accept formal responsibility for ensuring the integrity of the research process. They should foster an environment, a reward system, and a training process that encourage responsible research practices.[32]

The first version of the DFG code of practice, published in 1998, started with a general obligation to work *lege artis* (the state of the art).[33] The revised 2019 version contains 19 "guidelines" across three categories: "Principles," "Research Process," and "Noncompliance with Good Scientific Practice, Procedures." The guidelines on research process make up more than half of the text. The DFG Code exemplifies the spirit of scientific freedom (while acknowledging the need for restrictions to prevent abuse), a position that Heather Douglas advocated as recently as 2003 in her account of the moral responsibilities of scientists.[34] In her 2014 update, Douglas redefines "the moral terrain of science" in terms of forms of responsibility, and explicitly in relationship to society as well.[35]

In the course of the professionalization of science, attention inevitably also fell on the role of science in industry. C.F. von Weizsäcker directed his appeal to the state and the citizens, but today that would be insufficient. Many areas of research such as nuclear energy or space travel were previously reserved for government action due to their high investment requirements. Today, industry is taking on such tasks, to the extent that even the privately-conducted construction of nuclear weapons would be a legal rather than a technical–financial problem. The NAS Code of 1992 is primarily aimed at academic staff,[36] although it discusses its relevance for industry. Even if many areas of industry were fundamentally close to science, for example in the chemical and pharmaceutical sectors, the capitalist industrial exploitation of science takes place far from any elite responsibility or the corporate responsibility of science.

---

[29] Merton (1973), p. 276.

[30] Panel on Scientific Responsibility... (1992).

[31] DFG (2013/1998).

[32] NAS (1992), p. 13.

[33] This is in principle, an old legal version of the expectation of expert work (cf. Mieg, 2001/2012).

[34] Douglas (2003)

[35] Douglas (2014)

[36] NAS (1992), p. 23.

This made it necessary to critically consider *dual use*, i.e., research whose results can be used both civilly and militarily, or in an ethically questionable manner, e.g., to manipulate people. In this context, the DFG and the Leopoldina jointly published a paper in 2014 entitled "Scientific Freedom and Scientific Responsibility: Recommendations for Handling Security-Relevant Research."[37] The long title likely refers more to the uncertainty of dealing with the subject than to a thematically focused treatise. Lenk sharply criticizes the text, saying that it remains analytically unclear.[38] However, this is not a code of ethics, but an additional deliberation for institutional, government-funded research, beyond or in addition to the guidelines for good scientific practice (and, on p. 11, to "researchers in the industrial sector"). Nevertheless, the introduction of priority rules, as requested by Lenk,[39] would have improved the text. Such rules clarify cases of conflict, for example: When does good scientific practice apply; when do general ethical considerations apply?

In the United Kingdom, the government had already attempted to develop a universal minimum code of conduct for scientists prior to 2007.[40] This comprised three points: *rigor* (lege artis), *respect* (respect for the law, respect for the freedom of decision of all those involved…), and *responsibility* (here, essentially: dialogue with society, responsibility towards society). However, professionalization means strengthening or at least demonstrating self-control in order to avoid alternative forms of control being imposed by the state. The DFG/Leopoldina text is to be seen precisely in this light of ensuring professional self-regulation. It was deemed necessary to proactively draw up internal regulations (on dual use technologies) within the profession before the state attempted to introduce its own regulations.

The Threefold Concept of Responsibility and Its Normative Unity

We can assume that the appeal for responsibility is always made when uncertainty and risks are involved and where everyday rules and law do not apply. I have always understood responsibility as relational, i.e., following the literal meaning: An

---

[37] The DFG publication is based—often with the same wording—on the "Guidelines and Rules of the Max Planck Society on a Responsible Approach to Freedom of Research and Research Risks," drawn up in 2010 (MPG, 2017). The US National Academy of Sciences, together with other associations, published a report on dual use in 2011, based on a workshop held in 2009 ("Challenges and Opportunities for Education About Dual Use. Issues in the Life Sciences"). Here the focus is on the potential (mis)application of academic training to biological weapons. The question of education plays a subordinate role in the DFG publication (point 7, p. 15).

[38] Cf. Lenk (in this volume, Ch. 2).

[39] Cf. Lenk (in this volume, Ch. 2).

[40] Cf. Government Office for Science (2007).

individual has to answer to someone for something.[41] Furthermore, the concept of responsibility can be explained in three meanings, which essentially belong together.[42] Firstly, in *retrospect,* responsibility is about guilt or merit and the attribution of the consequences of events, whether positive or negative. Responsibility, in retrospect, becomes relevant, for example, when a loss has occurred and a culprit is sought. In science, this ranges from cases of (falsified) data to the big questions of responsibility for deaths caused by accidents and wars (poison gas, nuclear and neutron bombs…) Secondly, *prospectively,* it is about assuming responsibility, i.e., a social service to be provided (responsibility as performance).[43] Hans Jonas has emphasized this aspect of responsibility. Science as a profession bears responsibility for systematic knowledge acquisition. Thirdly, responsibility can be explained in *terms of authority and coordination:* Someone is in charge by virtue of office, mandate, or role. The coordinative function is usually associated with *status.* Without this status-effective coordination function, it would be incomprehensible why responsibility can serve as a payment equivalent. Why should someone choose more responsibility rather than higher salary, when responsibility would mean performance in the first place? The three directions of meaning *practically* belong together, but can decouple at any time. Weber, in his ethics of responsibility, calls for the unity of responsibility[44]: Whoever assumes a political office (status responsibility) should act accordingly (responsibility as performance, prospective) and be responsible for the consequences of his actions (retrospective). The unity of responsibility is normative.

A further differentiation results from the institutional question. Do institutions have their own responsibility, or only the individuals involved? Politics and law know institutional or *corporate responsibility.*[45] Institutions can be entrusted with task responsibility, and if they perform poorly they can be dissolved. Commercial enterprises can be punished, e.g., for violating antitrust laws. With regard to the responsibility of individuals in relation to institutions, we should distinguish between collective and co-responsibility. *Co-responsibility* corresponds to the principle of agreed complicity and refers to the original, individual decision to join an institution. In the case of professionalized science, the original decision means to follow a path as a professional scientist. Then we share responsibility for science in society. It may be a matter of further developing the organizational structures of science or communicating research results appropriately. In the case of *collective responsibility,* such a decision of origin is not essential or is only conceivable *ex negativo* via opting out. States are sometimes held responsible by other states, with

---

[41] The relational concept of responsibility is now standard in German. See Mieg (2015) and Lenk (in this volume, p. 48).

[42] See Mieg (1994a, 2015).

[43] See Mieg (1994a, b).

[44] Cf. Weber (1919/2004).

[45] See French (1984, 1992).

often drastic consequences for the individual. As we are citizens of a state by birth and law, it is not easy to deselect this status.

Established responsibility systems are *value neutral*. Already Cicero discussed that criminal gangs may very well develop an internal morality, i.e., clarify responsibilities, obligations, and reward or punish accordingly; this is documented as far back as the days of piracy.[46] Nevertheless, they remain criminals. The presumed freedom of values is not surprising if we consider the three meanings of responsibility: *ex post*—mostly negative (Who was to blame?)[47]; *ex ante*—mostly uncomfortably appellative (This person should show greater responsibility!); with regard to status—positive (She is the boss). Max Weber introduced the ethics of responsibility in contrast to an ethics of conviction. Acting in accordance with the ethics of responsibility takes into account the possible consequences of one's own actions, anticipating *conflicts of values*,[48] whereas from the point of view of an ethics of conviction only the orientation towards a canon of values counts. In contrast to Weber, C.F. von Weizsäcker bases his concept of responsibility, and thus the responsibility of science, on religion.[49]

This sober view of responsibility places institutions in the spotlight. There is a differentiated division of labor and thus responsibility in science, between scientific institutions (e.g., DFG or IPCC) on the one hand and individual scientists and researchers on the other. It is *ultra posse nemo obligatur*, institutions can achieve more and different things than a single person, we can expect different things from them. As a global organization, the IPCC lends climate change research a political and scientific weight that individual scientists, however well respected, would never achieve. Professionalization has given science a new and different power. No longer as a 'state religion' but as a knowledge method administrator, a global service provider. For this reason, science must expose itself to public criticism.

Science and Values: The New Social Responsibility of Science

In November 1917 Max Weber gave a lecture on science as a vocation. Some of the problems and phenomena that Weber addressed are likely to continue to determine the existence of scientists, for example, the necessary specialization and fixation on a narrow field of research, which is the only way to make progress in science; or the "double face" of science at universities, namely the combination of teaching on the one hand and research on the other. If one reads Weber's account through the eyes of a present-day researcher, it seems somewhat antiquated. This impression is due

---

[46] For example, just behavior among robbers, see Cicero (De officiis, 2nd book, section 11).

[47] French (1992) noted that for a while the Anglo-American discussion of responsibility revolved solely around responsibility barter games and the avoidance of responsibility (p. 2).

[48] Cf. Starr (1999).

[49] Cf. Liebert (2011).

above all to Weber's continual appeals against the presence of demagogues and prophets at universities and the attempts to distinguish between science and theology. Demagogues and theology still exist today, but do not currently present a problem; in Weber's time, the university had not yet freed itself from authoritarian state structures and officials, whose reasoning was perceived within the universities as mediocre because it comprised unreflective indoctrination. In the meantime, the universities have managed to drive out the authoritarian state. Science as a profession has long since emerged from the status of a civilized intellectual aristocracy. Despite all its precariousness, science has become a *plannable pattern for personal life plans*, from a merely "inner profession" (Weber) to a privileged, "outer" profession[50] whose freedom is anchored in law. Unlike in Weber's time, there is no longer any need to argue about freedom from value in science.

Moreover, the new professionalized role allows science to expand into value-defined areas. There have always been committed researchers who carried out value-oriented research, for example in community or peace research. However, among other colleagues, this was often perceived as bad research and part-time work in the sense of do-gooding. That particular situation changed with the advent of *environmental sciences*, applying scientific processes to environmental issues. The field is clearly not only about understanding and explaining environmental processes, but always about environmental protection. Environmental science requires a rethink, because:

- Environmental problems do not stop at disciplinary borders, and hence force different disciplines to work together;
- The dynamics of local environmental systems cannot be understood without the inclusion of *local system experts*.[51]

It was in this spirit that the Environmental Sciences program was established at ETH Zurich at the end of the 1980s. Wolf Lepenies, then Rector of the Wissenschaftskolleg in Berlin, praised the program as an example of "committed scholarship" and the "return of values to science"[52]:

> Vigilance and a sense of value oriented towards the public spirit are also awakened in this course of study by the fact that questions from the humanities and social sciences, economics and jurisprudence have been part of the curriculum for natural scientists from the very beginning. In this way, attention to the social embedding of one's own research and its possible social consequences is achieved without any pathos: it is not part of a facultative Studium generale, but is a natural part of everyday scientific life.[53]

---

[50] The antiquated impression also stems from many things that Weber did not address, such as fraud in science, the role of peer review, and performance evaluation (impact factors, teaching).

[51] Mieg (2001).

[52] Lepenies (1997), p. 44 [translated].

[53] Op. cit.

It became clear that environmental science cannot be pursued without "*co-production of knowledge*" (Jasanoff)[54] or "*mutual learning*" by science and society (R.W. Scholz).[55] The reasons are epistemic in nature. Knowledge is not otherwise available. This led to citizen science not only in birdwatching but also in "transdisciplinary" urban and regional development projects.[56] New disciplines such as *Sustainability Science* or *Global Environmental System Science* have emerged.[57] There, sustainability (as a value) has become a general obligation of scientific research.[58] Accordingly, the civic responsibility of scientists and scholars in these subjects has increased.

Today, science must present its contribution to society in order to justify the large, ongoing public investments. The question of application is as old as science and universities. However, expansion and professionalization have fueled the discussion of the *social responsibility* of science. The sciences share the fate of other, later professions, such as architecture, business management or social work, whose performance is seen by some as an unnecessary luxury and which are required to legitimize themselves anew through social responsibility—for example, through community orientation or a commitment to sustainable development or social equality[59]. In 1992, the U.S. National Academy of Sciences, along with partner institutions, published a manifesto on "responsible science." It clearly states that science has a responsibility to a *larger community*:

> Because scientists and the achievements of science have earned the respect of society at large, the behavior of scientists must accord not only with the expectations of scientific colleagues, but also with those of a larger community.[60]

Computer-Aided Simulation as a New Method of Knowledge—New Relationship to Politics

The upswing in environmental science and especially in sustainability studies would not be possible without a new methodological confidence. Especially thanks to the availability of computers, unexpected possibilities for modeling and simulation

---

[54] See Jasanoff (2004).

[55] See, e.g., Scholz et al. (2000).

[56] Op. cit.

[57] Often under the general title of "transformative" science (e.g., Schneidewind & Singer-Brodowski, 2014).

[58] The new discipline liked to see their members as "environmental physicians" concerned about the health of nature (see Mieg & Frischknecht, 2014). Somewhat more appropriate would have been to compare environmental science with architecture. Building is complex, does not end at the borders of disciplines, and requires the practical cooperation of many actors: client, trades, accessories, specialist planning…

[59] For business management cf. Valentine and Fleischman (2008).

[60] Panel on Scientific Responsibility… (1992), p. 7.

have arisen. Models simulate and depict the world or parts of it. Creating a model, in the sense of a simple simulation, was long considered scientifically insufficient.[61] Classical methods involve data collection, from simple observation to complex measurements, and experimentation. In the experiment, influencing variables are systematically varied and the resulting changes measured, ultimately to test causal relationships. Models become interesting if they can be used to test hypotheses and clarify specific questions. Climate change research is unthinkable without simulation.[62] Modelling has long been known from macroeconomics and has sometimes been ridiculed. In the meantime, computer-aided simulation has been established as a knowledge method.[63] In climate change research, several models based on different approaches are usually used. The resulting scenarios, i.e., possible world conditions, can then be evaluated: Is the development described economically viable? Is it socially just?

With professionalization, science has also gained a new role in relation to politics. The redefinition became necessary at the latest since the oil crisis of 1973. Science had not foreseen this and therefore did not offer a solution, especially in the political dimension of the crisis. Scientific policy advice was largely based on decision-theoretical models and approaches.[64] In comparison, mathematical game theory was embryonic. Moreover, science had predicted neither the fall of the Berlin Wall nor the financial crisis of 2007/08 and thus ran into a problem of legitimacy. On the other hand, scientifically based input is essential for national administrations, e.g., statistical offices or in health monitoring. Politicians demand certainty from science; anything else is often dismissed as a question of faith or opinion-forming. In science communication, it remains difficult to communicate probabilities and present risks. Simulation allows us to deal with hypothetical knowledge and risks, as we see what happens if we do not know... This approach is increasingly proving its worth in local infrastructure planning and global climate change research, for which the IPCC stands.[65]

The inclusion of values, for example for the purpose of environmental protection, together with new forms of knowledge production, e.g., the co-production of

---

[61] Simulation was long frowned upon by many scientists; only the classical experiment counted. A Leibniz Prize winner in the field of psychology told me that as the editor of a major academic journal he had to reject an article as being unscientific for the sole reason that it was based entirely on simulation. I knew the contribution, and found it sufficiently interesting for scientific discussion. This was in the 1990s, even before the great replication crisis in psychology (in which many of the classical experimental results proved to be unreproducible).

[62] See, e.g., Girod et al. (2009).

[63] On simulation as a path of knowledge, see Mieg (2019). This paper represents the latest attempt to define forms of research—independent of or "across" disciplines. The starting point was a study by the German Science Council (Wissenschaftsrat, 2012).

[64] See Mintzberg (1994).

[65] In the case of climate change research, it can be shown that scientific modelling—with the loss of 'best practice' scenarios—is compatible with policy (Mieg, 2004; Girod et al., 2009). The IPCC (Intergovernmental Panel on Climate Change) provides the simulations and leaves the option-finding to policy makers.

knowledge, open up new governance options for science policy. This is reflected at EU level in the normative framework of *Responsible Research and Innovation* (RRI). Here, two values or principles—responsibility and innovation—are combined in a trade-off between protection and change, a difficult task in which science must now prove its potential social contribution. The classic definition of RRI, which we find in von Schomberg, is correspondingly pragmatic:

> Responsible Research and Innovation is a transparent, interactive process by which societal actors and innovators become mutually responsive to each other with a view to the (ethical) acceptability, sustainability and societal desirability of the innovation process and its marketable products (in order to allow a proper embedding of scientific and technological advances in our society).[66]

In practical terms, RRI means that research projects must provide for specific measures, namely with regard to *public engagement, open access, gender, ethics,* and *science education.*[67] RRI had been actively promoted by the European Commission as a cross-cutting issue in its Horizon2020 funding scheme (2014–2020) and defined the sub-programme "Science with and for Society" (SwafS).[68]

Consequences of Professionalization, Conclusion

Science as a profession has a legitimate self-interest. Individual scientists must be able to live from their scientific work; scientific institutions rely on state subsidies. Thus, over time, *operating conditions of science* have emerged. In addition to peer review, these are the *disciplines* that organize the scientific process.[69] The expansion of science as a profession brings with it new professional roles, e.g., research management, and new responsibilities in science, which cross the systemic boundaries of science both inside and outside. The task now is to create structures, provide resources, develop standards… Science is no longer alone in the scholarly republic. Academic administration positions are being created and, in order to ensure quality, these must be made career-ready.[70]

Another consequence of professionalization, which is reluctantly perceived, is a *new relativization of science.* After its adoption as a (potential) state religion, i.e., 'scientism,' science is today only one voice among many. New and unexpected is

---

[66] Von Schomberg (2013), p. 19; cf. Macnagthen (in this volume, Ch. 5).

[67] Cf.   https://ec.europa.eu/programmes/horizon2020/en/h2020-section/responsible-research-innovation

[68] European Commission. (2020); cf. Oevermann et al. (in this volume, Ch.12).

[69] Cf. for example Mieg and Evetts (2018).

[70] The term *Third Space* has become established for these new academic professional roles, which arise mainly at universities (quality management, administration of study matters, etc.) and which require qualification via a university degree. From the perspective of professional research, one could also speak of academic semi-professions. In the academic field such a term is forbidden simply because of the ethos of anti-hierarchical collegiality.

the fight against disinformation, termed 'fake news.'[71] Here we can see a responsibility of science as a whole, patiently putting things right, adding question to things… Helmut F. Spinner[72] saw a parallel between journalism and science. However, journalism is subject to different operating conditions than science, although the noble goal of enlightenment may be the same.[73] Science communication is a task that falls within the responsibility of science as a profession, but does not necessarily have to be relevant to science training or research activities. For the majority of researchers, research is the main focus whereas communication is secondary. In addition to research, working science often does not even have time for anything other than writing contributions to scientific journals.[74]

Another consequence of professionalization (one could also speak of a creeping effect) is shown in the *power of formalization*. The practice is increasingly permeated by formulas, technical terms, and model approaches. Not as truths, but as working tools of scientifically trained people in practice, who bring their expertise and professional socialization to their work.[75] This formalization means a scientificization of practice. Compared to the situation in which C. F. von Weizsäcker saw himself and science, this professionalization brings a certain relief but also new burden for individual scientists. Science no longer exists in a vacuum without 'administrative stuff'; that is the price of professionalization. The weight of responsibility in science has been redistributed. Today, science communication and dialogue with society is more a focus of the science-supporting institutions, i.e., associations and universities. Individual scientists are largely free to make such commitments. On the other hand, they have a new, obligatory co-responsibility in the cooperation with the practice. It is important to ensure that science and the appropriate discipline are well represented and that science-based procedures are further developed in the relevant field of work.[76]

To conclude, science as a systematic method of knowldege production is much older than the science profession.[77] The responsibility of science as a system has

---

[71] Cf. Zimmermann (in this volume, Ch. 9).

[72] Spinner (1985).

[73] For example, science must never violate the principle of transparency, whereas in investigative journalism covert action may be necessary and professionally justifiable. The differences become significant when we distinguish the scientific from the journalistic interview (Mieg & Näf, 2006, p. 12).

[74] This is especially true for the natural sciences. When I showed my habilitation thesis, printed as a book, to a colleague from the Atmospheric Chemistry Department at ETH Zurich, he said, slightly astonished, that unfortunately he might only find time to write books once he became an emeritus (i.e., largely retired from everyday academic commitments).

[75] This was the surprising finding of my outreach study on the impact of environmental science in Switzerland (cf. Mieg et al. 2012): the "language" of environmental science was adopted in offices and associations, if only because their graduates worked there.

[76] Such "practice-developing research" can be considered a form of knowledge in its own right (cf. Mieg, 2019).

[77] This refers to science as a social functional system within the framework of a functionally differentiated society.

changed with professionalization. Here we can distinguish an inward and an outward responsibility (see Figure 2). The inward responsibility refers to good scientific practice; the outward responsibility to the social or organizational context, which can range from education and scientific sense-making to economic and war fitness, and today may include a contribution to civil society development, to demonstrate social responsibility.[78] With professionalization, science has received a tacit administrative mandate for systematic knowledge acquisition and emancipated itself from further-reaching appropriation. Many inner-scientific responsibilities, such as method development or quality assurance, have risen over time and passed into institutional hands, namely to the disciplines. In the past, science was embodied in the individual scientists and scholars, whereas today it is embodied in the disciplines. In summary, Fig. 4.2 provides an overview of the current scope of the responsibility of science.

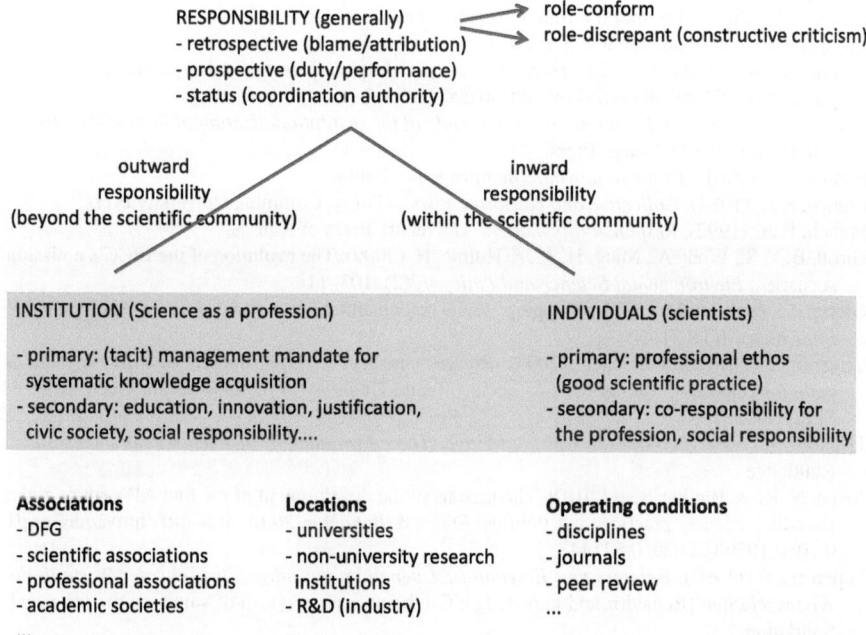

Fig. 4.2 Responsibility of professionalized science (shaded in gray): With the professionalization of science, individual scientists are relieved of many non-scientific constraints and can plan their careers and live from them (with similar professional risks as in an independent architectural practice). For more details cf. Mieg (2015) and Mieg and Evetts (2018)

---

[78] Cf. also Glerup and Horst (2014).

References

Abbott, A. (1988). *The system of professions*. The University of Chicago Press.
Bartosch, U., Litfin, G., Braun, R., & Neuneck, G. (Hrsg.). (2011). *Verantwortung von Wissenschaft und Forschung in einer globalisierten Welt* [Responsibility of science and research in a globalized world]. LIT.
Ben-David, J. (1972). The profession of science and its powers. *Minerva, 10*(3), 362–383.
Bundesministerium für Bildung und Forschung BMBF (2018). *Bundesbericht Bildung und Innovation 2018* [Federal Report Education and Innovation 2018]. BMBF.
Deutsche Forschungsgemeinschaft DFG. (2013). *Safeguarding good scientific practice: Memorandum*. Wiley-VCH. (First edition in 1998, online via DFG).
Deutsche Forschungsgemeinschaft DFG. (2019). *Leitlinien zur Sicherung guter wissenschaftlicher Praxis: Kodex*. DFG.
Deutsche Forschungsgemeinschaft DFG und Deutsche Akademie der Naturforscher Leopoldina e.V. (2014). *Wissenschaftsfreiheit und Wissenschaftsverantwortung: Empfehlungen zum Umgang mit sicherheitsrelevanter Forschung. Scientific freedom and scientific responsibility: Recommendations for handling security-relevant research*. DFG und Leopoldina.
Douglas, H. (2003). The moral responsibilities of scientists (tensions between autonomy and responsibility). *American Philosophical Quarterly, 40*(1), 59–68.
Douglas, H. (2014). The moral terrain of science. *Erkenntnis, 79*(5), 961–979.
European Commission. (2020). *Science with and for Society in Horizon 2020: Achievements and recommendations for Horizon Europe*. https://op.europa.eu/en/publication-detail/-/publication/770d9270-cbc7-11ea-adf7-01aa75ed7
Freidson, E. (1986). *Professional powers: A study of the institutionalization of formal knowledge*. The University of Chicago Press.
Freidson, E. (2001). *Professionalism: The third logic*. Polity.
French, P. A. (1984). *Collective and corporate responsibility*. Columbia University Press.
French, P. A. (1992). *Responsibility matters*. University Press of Kansas.
Girod, B. V. S., Wiek, A., Mieg, H. A., & Hulme, H. (2009). The evolution of the IPCC's emission scenarios. *Environmental Science and Policy, 12*(2), 103–118.
Glerup, C., & Horst, M. (2014). Mapping 'social responsibility' in science. *Journal of Responsible Innovation, 1*(1), 31–50.
Government Office for Science. (2007). *Rigour – respect – responsibility: A universal ethical code for scientists*. The Science & Society Team, Department for Innovation, Universities & Skills (DIUS).
Jasanoff, S. (Ed.). (2004). *States of knowledge: The co-production of science and social order*. Routledge.
Joyce, N. R., & Rankin, T. J. (2010). The lessons of the development of the first APA ethics code: Blending science, practice, and Politics. *Ethics & Behavior, 20*(6), 466–481. https://doi.org/1 0.1080/10508422.2010.521448
Lepenies, W. (1997). B*enimm und Erkenntnis: Über die notwendige Rückkehr der Werte in die Wissenschaften* [Behavior and knowledge: On the necessary return of values to the sciences]. Suhrkamp.
Liebert, W. (2011). Carl Friedrich von Weizsäcker zur Verantwortung der Wissenschaft [Carl Friedrich von Weizsäcker on the responsibility of science]. In U. Bartosch, G. Litfin, R. Braun, & G. Neuneck (Eds.), *Verantwortung von Wissenschaft und Forschung in einer globalisierten Welt* (pp. S. 273–S. 286). LIT.
Max-Planck-Gesellschaft zur Förderung der Wissenschaften e.V. MPG. (2017). *Guidelines and Rules of the Max Planck Society on a Responsible Approach to Freedom of Research and Research Risks*. MPG.

Merton, R. K. (1973/1942). The normative structure of science. In N. W. Storer (Ed.), *The sociology of science* (pp. 267-278). University of Chicago Press.. (Original work appeared in 1942)

Mieg, H. A. (1994a). Verantwortung als Leistung [Responsibility as performance]. *Zeitschrift für Sozialpsychologie, 25*, 208–216.

Mieg, H. A. (1994b). *Verantwortung: Moralische Motivation und die Bewältigung sozialer Komplexität* [Responsibility: Moral motivation and coping with social complexity]. Westdeutscher Verlag.

Mieg, H. A. (2001). *The social psychology of expertise: Case studies in research, professional domains, and expert roles*. Lawrence Erlbaum Associates.

Mieg, H. A. (2004). The precarious role of scenarios in global environmental politics: Political options versus scientific projections. In F. Biermann, S. Campe, & K. Jacob (Eds.), *Knowledge for the Sustainability Transition: The Challenge for Social Science*. Global Governance Project. (online auf academia.edu).

Mieg, H. A. (2005). Professionalisierung. In F. Rauner (Ed.), *Handbuch der Berufsbildungsforschung* (pp. S. 342–S. 349). Bertelsmann. (Neuauflage 2018).

Mieg, H. A. (2015). Social reflection, performed role-conformant and role-discrepant responsibility, and the unity of responsibility: A social psychological perspective. *Soziale Systeme, 19*(2), 259–281.

Mieg, H. A. (2018). Professionalisierung – eine konzeptionelle Wiederbelebung [Professionalization – a conceptual revival]. In H. A. Mieg, *Professionalisierung: Essays zu Expertentum, Verberuflichung und professionellem Handeln* (Einleitung, S. 11-36). Verlag der Fachhochschule Potsdam.

Mieg, H. A. (2019). Forms of research within strategies for implementing undergraduate research. *ZFHE, 14*(1), 79–94.

Mieg, H. A., Hansmann, R., & Frischknecht, P. M. (2012). National sustainability outreach assessment based on human and social capital: The case of Environmental Sciences in Switzerland. *Sustainability, 4*(1), 17–41.

Mieg, H. A., & Evetts, J. (2018). Professionalism, science, and expert roles: A social perspective. In K. A. Ericsson, R. R. Hoffman, A. Kozbelt, & A. M. Williams (Eds.), *The Cambridge handbook of expertise and expert performance* (2nd ed., pp. 127–148). Cambridge University Press.

Mieg, H. A., & Frischknecht, P. M. (2014). Multidisziplinär, antidisziplinär, disziplinär? Die Geschichte der Umweltnaturwissenschaften an der ETH Zürich [Multidisciplinary, antidisciplinary, disciplinary? The history of environmental sciences at ETH Zurich]. In B. Engler (Ed.), *Disziplin/Discipline* (pp. S. 135–S. 169). Fribourg Academic Press.

Mieg, H. A., & Näf, M. (2006). *Experteninterviews in den Umwelt- und Planungswissenschaften: Eine Einführung und Anleitung* [Expert interviews in environmental and planning sciences: An introduction and guidance]. Pabst.

Mintzberg, H. (1994). *The rise and fall of strategic planning*. Free Press.

NAS National Academy of Sciences (1992). *Responsible science, Volume I: Ensuring the integrity of the research process*. National Academies Press.

Oevermann, U. (1996). Theoretische Skizze einer revidierten Theorie professionalisierten Handelns [Theoretical sketch of a revised theory of professionalized action]. In A. Combe & W. Helsper (Eds.), *Pädagogische Professionalität: Untersuchungen zum Typus pädagogischen Handelns* (pp. 70–182). Suhrkamp.

Oevermann, U. (2005). Wissenschaft als Beruf: Die Professionalisierung wissenschaftlichen Handelns und die gegenwärtige Universitätsentwicklung [Science as a profession: The professionalization of scientific activities and the current development of universities]. *die hochschule, 14*(1), 15–51.

Panel on Scientific Responsibility and the Conduct of Research; Committee on Science, Engineering, and Public Policy; National Academy of Sciences, National Academy of Engineering, & Institute of Medicine. (1992). *Responsible science, Volume I: Ensuring the integrity of the research process*. National Academy Press.

Schneidewind, U., & Singer-Brodowski, M. (2014). *Transformative Wissenschaft. Klimawandel im deutschen Wissenschafts- und Hochschulsystem* (2nd, rev. ed.) [*Transformative science. Climate change in the German science and higher education system*]. Metropolis-Verlag.

Scholz, R. W., Mieg, H. A., & Oswald, J. (2000). Transdisciplinarity in groundwater management: Towards mutual learning of science and society. *Water, Air, & Soil Pollution, 123*, 477–487.

Schurr, G. M. (1982). Toward a code of ethics for academics. *The Journal of Higher Education, 53*(3), 318–334.

Spinner, H. F. (1985). *Das "wissenschaftliche Ethos" als Sonderethik des Wissens [The "scientific ethos" as a special ethic of knowledge]*. Mohr.

Stark, L. (2010). The science of ethics: Deception, the resilient self, and the APA code of ethics, 1966–1973. *Journal of the History of the Behavioral Sciences, 46*(4), 337–370.

Starr, B. E. (1999). The structure of Max Weber's ethic of responsibility. *Journal of Religious Ethics, 27*(3), 407–434.

Stichweh, R. (1994). *Wissenschaft, Universität, Professionen* [Science, university, professions]. Suhrkamp.

Stichweh, R. (1996). Professionen in einer funktional differenzierten Gesellschaft [Professions in a functionally differentiated society]. In A. Combe & W. Helsper (Eds.), *Pädagogische Professionalität. Untersuchungen zum Typus pädagogischen Handelns* (pp. S. 49–S. 69). Suhrkamp.

The National Academies of Sciences, Engineering, Medicince. (2011). *Challenges and opportunities for education about dual use: Issues in the life sciences*. National Academies Press.

Valentine, S., & Fleischman, G. (2008). Professional ethical standards, corporate social responsibility, and the perceived role of ethics and social responsibility. *Journal of Business Ethics, 82*(3), 657–666.

von Schomberg, R. (2013). A vision of Responsible Research and Innovation. In R. Owen, J. Bessant, & M. Heintz (Eds.), *Responsible innovation: Managing the responsible emergence of science and innovation in society* (pp. 51–74). Wiley.

von Weizsäcker, C. F. (2011). *Die Verantwortung der Wissenschaft im Atomzeitalter* (7. ed.) [The responsibility of science in the nuclear age]. Vandenhoeck & Ruprecht.

von Weizsäcker, E.U. (1986). Die gedoppelte Scientific Community [The doubled scientific community]. In U. Bartosch, G. Litfin, R. Braun & G. Neuneck (Hrsg.). (2011). *Verantwortung von Wissenschaft und Forschung in einer globalisierten Welt* (S. 221-S. 230). LIT.

Weber, M. (1919/2004). *Science as a Vocation* (edited by D. Owen & T. B. Strong, translated by R. Livingstone). Indianapolis: Hackett.

Wissenschaftsrat. (2012). *Empfehlungen zur Weiterentwicklung der wissenschaftlichen Informationsinfrastrukturen in Deutschland bis 2020* [Recommendations for the further development of scientific information infrastructures in Germany until 2020]. Wissenschaftsrat. (Online)

Part II
Insights Into the Quest for Responsibility in Science–Society Interactions

In this part of the book, we look at the two sides of the interface between science and society. On the one hand, there is the issue of how science can be controlled. On the other hand, the concern of science is to ensure its own independence. This part includes four analytical chapters. Two chapters deal with current regulatory issues at the EU level, on the one hand for the purpose of research policy (RRI, Responsible Research and Innovation), and on the other hand with a view to implementing the Precautionary Principle in European legislation. Two other chapters examine the operation of science, firstly at moral questions linked to the scientific discovery of nuclear energy in the 1930s, and secondly at the mechanisms of institutionalization and interdisciplinarity that help to consolidate as well as advance science.

Chapter 5: Phil Macnaghten's contribution, "From the Linear Model to Responsible Research and Innovation (RRI)," is the prelude to this second part of our volume, concerned with clarifying responsibilities at the interfaces between science and society. From a science policy perspective, RRI provides a development and governance framework for science, favored especially by EU research policy.

Chapter 6: Horst Kant's contribution, "Research and Use of Nuclear Energy—Its Ambivalence(s) in Historical Context," discusses the discovery of nuclear fission in the late 1930s. From the perspective of the history of science, it becomes apparent that many physicists involved in early atomic research were aware of the political–military potential, and that some sought to prevent "misuse" of atomic energy (especially in the 1950s).

Chapter 7: Heinrich Parthey's paper, "Institutionalization, Interdisciplinarity, and Ambivalence in Research Situations," argues from the perspective of science studies that science's strength lies in institutionalization, while ambivalence of scientific research can lead to ethical issues.

Chapter 8: In their contribution "The Application of the Precautionary Principle in the EU," Kristel de Smedt & Ellen Vos introduce the Precautionary Principle, which is intended to make the consequences of technological innovations controllable. From a legal point of view, the application of the Precautionary Principle is problematic.

About the authors: *Phil (Philip Martin) Macnaghten* (born 1965) teaches at Wageningen University & Research, the Netherlands. He is one of the scholars who have become authoritative for RRI. *Horst Kant* (born 1946), physicist and historian of science, has worked both for the Academy of Sciences of the former GDR and, most recently, for the Max Planck Institute for the History of Science, Berlin. *Heinrich Parthey* (1936–2020) was a charismatic philosopher of science and became founder of the Berlin Society for the Study of Science, which he led until 2020. *Ellen Vos* teaches law at Maastricht University, the Netherlands. *Kristel de Smedt* teaches at Maastricht University and Hasselt University, Belgium. They lead the EU project RECIPES on the role of the Precautionary Principle (https://recipes-project.eu).

Chapter 5
Models of Science Policy: From the Linear Model to Responsible Research and Innovation

Phil Macnaghten

Abstract In this paper I discuss four different paradigms through which science and technology have been governed, situating each in historical context. Starting with the ubiquitous 'linear model of innovation' I locate its origins and provenance, how it came to be replaced, at least in part, through a 'grand challenges' paradigm of science policy and funding; how this paradigm in turn has been subjected to rigorous analytical critique by a co-production model of science and society, and how it is being put into practice, in part, through a framework of responsible research and innovation.

The Linear Model of Innovation

The Second World War and its immediate aftermath signalled a critical moment in the unfolding relationship between science, society and the state, especially in the United States. The Manhattan Project in particular, involving the coordination of infrastructure and personnel in the development and production of the US nuclear programme, had demonstrated the utility of science in public policy, in this case its role in helping to win the war through the detonation of two atomic bombs over Japan. In November 1944, President Roosevelt commissioned Vannevar Bush, who played a formative role in administrating wartime military R&D through heading the US Office of Scientific Research and Development (OSRD), to produce a report laying out the contributions of science to the war effort, and their wider implications for future governmental funding of science. What emerged in July 1945 was the report, *Science – The Endless Frontier* (Bush, 1945), that became the hallmark of American policy in science and technology, and the blueprint and justification for many decades of increased funding in American science.

The Bush report is associated with the linear model of innovation, postulating that the knowledge creation and application process starts with (the government

P. Macnaghten (✉)
Wageningen University, Wageningen, The Netherlands
e-mail: philip.macnaghten@wur.nl

© The Author(s) 2022   93
H. A. Mieg (ed.), *The Responsibility of Science*, Studies in History and
Philosophy of Science 57, https://doi.org/10.1007/978-3-030-91597-1_5

funding of) basic research, which then leads to applied research and development, culminating with production and diffusion, and associated societal benefit. Even if this sequential linkage may have been added post hoc, partially and imperfectly reflected in Bush's actual report (see Edgerton, 2004), it nevertheless developed an iconic status as the origin and source of a dominant science policy narrative in which pure curiosity-driven science (knowledge for its own sake) was seen as both opposed to and superior to applied science, effectively operating as the seed from which applied research emerges, the economy grows and society prospers (Godin, 2006). As Jasanoff (2003) argues, the metaphor that gripped the policy imagination was the pipeline: 'With technological innovation commanding huge rewards in the marketplace, market considerations were deemed sufficient to drive science through the pipeline of research and development into commercialisation' (Jasanoff, 2003: 228). This logic was given further impetus by the diffusion of innovation literature, notably in E.M. Rogers' classic text (1962), which again adopted a linear and determinist model of science-based innovation diffusing into society with beneficial consequences.

Central to the post-WW2 science policy narrative was the concept of the social contract, namely that in exchange for the provision of funds, scientists—with sufficient autonomy and minimal interference—would provide authoritative and practical knowledge that would be seamlessly turned into development and commercialisation. The linear model understands science and policy as two separate spheres and activities. The responsibility of scientists is first and foremost to conduct good science, typically seen as guaranteed by scientists and scientific institutions upholding and promoting the norms of communalism, universalism, disinterestedness and organised scepticism (Merton, 1973). The ideal of science was represented as 'The Republic of Science' (Polanyi, 1962), of science as separate from society and as a privileged site of knowledge production. The cardinal responsibility of science according to this model was primarily to safeguard the integrity and autonomy of science, not least through practices of peer review as the mechanism that guarantees the authority of science in making authoritative claims to truth, and thus ensuring its separation from the sphere of policy and politics.

This division of powers served the interests of both actors: for scientists, a steady and often growing income steam as well as considerable self-autonomy, while for politicians and policy-makers, a narrative in which they can claim that their policies are grounded in hard and objective evidence ('sound science') and not in subjective values or ideology. This division was also written into institutional arrangements for science policy. The Haldane Principle for instance, that the decision-making powers about what and how to spend research funds should be made by researchers rather than politicians, was written into national science funding bodies in the UK as far back as 1918, operating especially following the Second World War as a powerful narrative for self-regulation and for safeguarding the autonomy of science.

So far, we have described the linear model of science and technology, the assumptions that underpin its governance, including its optimistic and deterministic view of the relationship between pure science and social progress. Yet, as the twentieth century progressed, this model came increasingly to be under strain as providing robust governance in the face of real-world harms that derived from scientific and techno-logical innovation. Traditional notions of responsibility in science were that of safe-guarding scientific integrity, whereas responsibility in scientific governance came to include responsibility for impacts that were later found to be harmful to human health or the environment. The initial governance response was to acknowledge that (even well-conducted) science and technology could generate harms, but that these could be evaluated in advance, and within the bounds of scientific rationality, through practices of risk assessment. Following a report from the US National Research Council (1983), systematising the process of risk assessment for govern-ment agencies through the adoption of a formalised analytical framework, a rigor-ous and linear scheme was promoted and disseminated in which each step was based on available scientific evidence and in advance of the development of policy options. Risk assessment was thus a response to the problems of the linear model, but still very much within the linear model's framing and worldview.

Notwithstanding the efficacy of risk assessment to mitigate the harms associated with science and technology, notably in relation to chemicals and instances of pol-lution, it did little to anticipate or mitigate a number of high-profile technology disasters that took place throughout the latter half of the twentieth century, and that demonstrated that science and technology could produce large-scale (and possibly systemic) 'bads' that evaded the technical calculus of science-based risk assessment (Perrow, 1984). High-profile disasters ranged from the Three Mile Island nuclear accident in the United States in 1979, to the Bhopal Union Carbide gas disaster in India in 1984, the Chernobyl nuclear disaster in Ukraine in 1986, the 'mad cow' BSE controversy in the UK and Europe throughout the late 1980s and 1990s, and the GM food and crop controversy in the 1990s and 2000s first in Europe and then across much of the Global South. The nuclear issue in particular became a focal point throughout the 1970s and 1980s for wider concerns about technological modernity, manifested in large social movements mobilised against the potential of science-led innovation to produce cumulative unknown and potentially cataclysmic risks. Theorised most famously by the sociologist Ulrich Beck and his notion of modernity having entered into a new phase dubbed the risk society, science and technology were seen as having produced a new set of global risks that were unlim-ited in time and space, manufactured (rather than as acts of God), potentially irre-versible, incalculable, uninsurable, difficult or impossible to attribute, dependent on expert systems and institutions for their governance, and where society operated as an experiment in determining outcomes (Beck, 1992).

The saga of bovine spongiform encephalopathy (BSE) or 'mad cow' disease in the UK and Europe is one such risk that was woefully and inadequately governed by a reliance on formal processes of science-based risk assessment, and where the political controversy derived from the inadequate handling of a new disease in cattle under conditions of scientific uncertainty and ignorance, and in the context of Britain's laissez-faire political culture. In this case, despite reassurances from gov-ernment ministers, claiming innocently to be following scientific advice that a trans-mission across the species barrier would be highly unlikely (following from the available risk assessments at the time that there was no evidence that proved such a transmission could take place), a deadly degenerative brain disease spread from cattle to humans, escalating to such proportions as to threaten the very cohesion of the European Union (Macnaghten & Urry, 1998).

More generally, risk assessment as a formal mechanism of scientific governance came under sustained criticism (for an extension of this argument, see Jasanoff, 2016). First, it embodies a tacit presumption in favour of change in assuming that innovations should be accepted in the absence of demonstrable harm. Second, it prioritises short-term safety considerations over long-term, cumulative and systemic impacts, including those on the environment and quality of life. Third, it prioritises a priori assumptions of economic benefits with limited space for public deliberation of those benefits and their distribution on society. Fourth, it restricts the scope of what is considered to be 'scientific' expertise, typically to a restricted set of disciplines, with limited scope for accessing the knowledge of ordinary citizens. And fifth, it ignores the values and deep-seated cultural presuppositions that underpin how risks are framed, including the legitimacy of alternative framings.

The Grand Challenge Model of Science for Society

While the linear model has been criticised for failing to account for the (especially systemic) risks associated with late modernity, the model has also come under sustained criticism as offering an inadequate account of how the innovation system is (or should be) structured and for what ends. Throughout the latter part of the twentieth century, science and innovation became increasingly integrated and intertwined. The knowledge production system moved from the rarefied sphere of elite universities, government institutes and industry labs into new sites and places that now included think tanks, interdisciplinary research centres, spin-off companies and consultancies. Knowledge itself became less disciplinary based and more bound by context and practical application. Traditional forms of quality control via peer-based systems became expanded to include new voices and actors adding additional criteria related to the societal and economic impact of research. Variously framed using new intellectual concepts that included 'Mode 2 knowledge' (Gibbons et al., 1994), 'post-normal science' (Funtowicz & Ravetz, 1993), 'strategic science' (Irvine & Martin, 1984) and the 'triple helix' (Etzkowitz & Leydesdorff, 2000), a new model of knowledge production emerged in which science came to be represented as the production of socially robust or relevant knowledge, alongside and often in conflict with its traditional representation as knowledge for its own sake. Interestingly, Mode 2 authors, in a later book, contextualised this transformation to accounts of societal change, particularly the Risk Society and the Knowledge Society, where 'society now speaks back to science' (Nowotny et al., 2001: 50; see also Hessels & van Lente, 2008).

One institutional response to critiques of the linear model has been the development of initiatives aimed at ensuring that science priorities and agenda-setting processes respond to the key societal challenges of today and tomorrow. The 'grand challenge' approach to science funding best illustrates this approach. Historical examples of grand challenges range from the prize offered by the British Parliament for the calculation of longitude in 1714 to President Kennedy's challenge in the 1960s of landing a man on the Moon and returning him safely to Earth. However, it was in the 2000s that the concept developed into a central organising trope in science policy, propelled inter alia by the Gates Foundation as a way of mobilising the international community of scientists to work towards predefined global goals (Brooks et al., 2009). In European science policy, the Lund Declaration in 2009 was a critical moment, which emphasised that European science and technology must seek sustainable solutions in areas such as global warming, energy, water and food, ageing societies, public health, pandemics and security.

More generally, the concept has been embedded across a wide array of funding initiatives that has included most recently the European Commission's Framework 8 Horizon 2020 programme (€80 billion of funding available over 7 years from 2014 to 2020), as a challenge-based approach that reflects both the policy priorities of the European Union and the public concerns of European citizens. Legitimated as responding to normative targets enshrined in Treaty agreements, these include goals on health and wellbeing, food security, energy, climate change, inclusive societies and security. It assumes, in other words, that science does not necessarily, when left to its own self-regulating logics and processes, respond to the challenges that we as a society collectively face. It needs some degree of steering, or shaping, on the part of science policy institutions, to ensure alignment. It is thus embedded in a discourse about the goals, outcomes and ends of research.

Over the last decade, the grand challenge concept has become deeply embedded in science policy institutions, as a central and organising concept that appeals to national and international funding bodies, philanthropic trusts, public and private think tanks and universities alike. It operates not only as an organising device for research calls but also as a way of organising research in research-conducting organisations, notably universities. At my university, for example, Wageningen University configures its core mission and responsibility in strategic documents (e.g., in annual reports, strategic plans, corporate brochure) as that of producing 'science for impact', principally through responding to global societal challenges of food security and a healthy living environment (Ludwig et al., 2018).

The grand challenge concept is clearly aligned to the 'impact' agenda, where researchers increasingly have to demonstrate impact (or pathways to impact) in research funding applications and evaluation exercises. These concepts help reconfigure the social contract for science such that, at least in part, the responsibility of science is to respond to the world's most pressing societal problems, while the responsibility of science policy institutions is configured as that of ensuring that the best minds are working on the world's most pressing problems (Brooks et al., 2009). Perhaps not surprisingly, these initiatives prove controversial within the scientific community, as for example witnessed in a backlash from the scientific community to an initiative from one of the UK research council's, the Engineering and Physical Science Research Council (EPSRC), plans to prioritise its funding for grants, studentships and fellowships according to national importance criteria (its 'shaping capability' initiative, see Jump, 2014).

Flink and Kaldewey (2018) add a further analytical layer. They produce a historically situated linguistic analysis of the 'grand challenge' science policy concept and
the ways in which it has replaced the earlier figure of the scientist prevalent in the
linear model of innovation. Since at least Vannevar Bush's report, *The Endless
Frontier* (1945), the dominant figure of the scientist was that of a lone individualist,
discovering the frontiers of knowledge through pioneering or frontier research at the
rock face of knowledge. However, while the ideal-type of this kind of scientist was
that of 'the risk-taking behavior of rugged competitive individualists pioneering
into the unknown' (Flink & Kaldewey, 2018: 16), the grand challenge concept configured a different kind of scientist. The grand challenge scientific endeavour still
remains competitive but has now become collective, even sports-like, in the ways in
which teams are presented as fighting to achieve a significant long-term goal, the
accomplishment of which will have significant societal impact. This tends to favour
the organisation of science in highly interdisciplinary and collaborative units, such
as has become the case in Systems Biology or Synthetic Biology. Yet, even though
grand challenges, by definition, are attempts to respond to society and to the public
interest, the choice and framing of the challenges themselves have tended to remain
those that have been chosen top-down by funding organisations (Calvert, 2013), and
in ways that often lend themselves to 'silver bullet' technological solutions (Brooks
et al., 2009). Nevertheless, the grand challenge concept can be seen as part of an
attempt to establish a new social contract for the public funding of science, and as
an important counterweight to the other dynamic that has impacted on the autonomy of science—namely, the relentless influence on economic drivers that has come
to dominate research policy agendas (National Council on Bioethics, 2012).

The Co-Production Model of Science and Society

If the 'grand challenge' science-policy model seeks to reconfigure the social contract of science such that its core value lies, not with the pursuit of pure knowledge
but in providing solutions to the world's most pressing problems, the co-production
model and approach seeks to reconfigure the social contract in another direction.
While the linear model views science as the motor of societal progress, and while
the grand challenge model views science as the provider of solutions for society, the
model of co-production views the spheres of science and social order as mutually
constitutive of each other.

Developed by Sheila Jasanoff and colleagues, and building on decades of scholarship in science and technology studies (STS), the co-production concept criticises
the idea of science as producing incontrovertible fact. As Jasanoff and Simmet
claim: 'Facts that are designed to persuade publics are co-produced along with the
forms of politics that people desire and practice' (2017: 752). This takes place in
deciding which facts (or truth claims) to focus on (which is seen as a normative
issue), in identifying in whose interests the facts are used to support (given that facts
are never seen as independent from values or indeed ideology), and in observing

that public facts are achievements, or what Jasanoff and Simmet call 'precious collective commodities, arrived at ... through painstaking deliberation on values and slow sifting of alternative interpretations based on relevant observations and arguments' (2017: 763).

There are three broad implications that derive from this approach. First, if the authority and durability of public facts depend, not on their status as indelible truths, but on the virtues and values that have been built into the ethos of science over time (e.g., through careful observation, transparency, open critique and reasoned argument), it follows that we need to give special attention precisely to these virtues, and to how these have been cultivated over time by institutional practice, as an important constituent of democratic governance. Or as Jasanoff and Simmet claim: 'building strong truth regimes requires equal attention to the building of institutions and norms' (2017: 764).

Second, if science and social order are co-produced, then it becomes incumbent on the research enterprise to examine precisely the relationship in practice between scientific knowledge production and social order as evinced in particular sites. Variously studying in depth the operation of scientific advisory bodies, technical risk assessments, public inquiries, legal processes and public controversies, science and technology studies (STS) scholars have identified both the values out of which science is conducted, including the interests it serves, as well as the ways in which these configurations can, over time, contribute to the formation of new meanings of life, citizenship and politics, or what more generally can be dubbed 'social ordering' (see, amongst many others, Jasanoff, 1990, 2004; Miller, 2004; Owens, 2015; Rose, 2006).

Third, if it is acknowledged that science and social order are co-produced, even if unwittingly through forms of practice (not least due to the continued prevalence of the fact–value distinction and the long reach of the linear model), the question arises as to what are the values that underpin the scientific knowledge-production system (and their associated cultures), and to what extent these align with broader societal values. Indeed, to what extent have the values and priorities tacitly embedded in scientific innovation been subjected to democratic negotiation and reflection? Or, perhaps more worryingly, to what extent are dominant scientific values reflective of those of incumbent interests that may be, perhaps unwittingly, closing down possibilities for different scientific pathways linked to alternative visions of the social good (Stirling, 2008, 2014). Responding to these questions, a line of research has emerged since the late 1990s, particularly prevalent in northern parts of Europe, aimed at early-stage public and societal participation in technoscientific processes as a means of fostering democratic processes in the development, approach and use of science and technology. Such initiatives, funded both by national funding bodies as well as by international bodies such as the European Commission, are typically aimed at improving relations between science and society and restoring legitimacy (e.g., see European Commission, 2007). In practice, they have been developed for reasons that include: the belief that they will help restore public trust in science, avoid future controversy, lead to socially robust innovation policy, and render scientific culture and praxis more socially accountable and reflexive (Irwin, 2006;

Macnaghten, 2010). Initiatives aimed at public engagement in science have become a mainstay in the development of potentially controversial technology, notably in the new genetics, and have even been institutionally embedded into the machinery of government in such initiatives that include the UK Sciencewise dialogues on science and technology (Macnaghten & Chilvers, 2014). In academia, they have contributed to institutional initiatives that include Harvard University's *Science and Democracy Network*, and to the sub-discipline of public engagement studies (Chilvers & Kearnes, 2016).

A Framework of Responsible Research and Innovation

The responsible research and innovation (RRI) concept represents the most recent attempt to bridge the science and society divide in science policy. Actively promoted by the European Commission as a cross-cutting issue in its Horizon 2020 funding Scheme (2014–2020), and embedded in its sub-programme 'Science with and for Society' (SwafS), RRI emerged as a concept designed both to address European (grand) societal challenges and as a way to 'make science more attractive, raise the appetite of society for innovation, and open up research and innovation activities; allowing all societal actors to work together during the whole research and innovation process in order to better align both the process and its outcomes with the values, needs and expectations of European society' (European Commission, 2013: 1). To some extent RRI has been a mere 'umbrella term', where RRI is operationalised through projects aimed at developing progress in traditional domains of European Commission activity, nominally in the so-called five keys of gender, ethics, open science, education to science, and the engagement of citizens and civil society in research and innovation activities (Rip, 2016). Under this interpretation RRI is simply a continuation of initiatives aimed at bringing society into EU research policy, starting with its Framework 6 programme (2002–2006) 'Science and Society' and its follow-on Framework 7 programme (2007–2013) 'Science in Society'; identified as a (yet another) top-down construct, introduced by policymakers and not by the research field itself (Zwart et al., 2014: 2), standing 'far from the real identity work of scientists' (Flink & Kaldewey, 2018: 18).

Yet, another—and potentially more transformative—articulation of the RRI concept is also available. Alongside colleagues Richard Owen and Jack Stilgoe, I have been involved in developing a framework of responsible innovation for the UK research councils. Our intention at the time was to develop a framework out of at least three decades of research in science and technology studies (STS), building on the co-production model as articulated above. Our starting point drew on the observation that from the mid- twentieth century onwards, as the power of science and technology to produce both benefit and harm had become clearer, it had become apparent that debates concerning *responsibility* in science need to be broadened to extend both to their collective and to their external impacts (foreseen and unforeseen) on society. This follows directly from the co-production model as articulated above.

Responsibility in science governance has historically been concerned with the 'products' of science and innovation, particularly impacts that are later found to be unacceptable or harmful to society or the environment. Recognition of the limitations of governance by market choice has led to the progressive introduction of post hoc—and often risk-based—regulation, such as in the regulation of chemicals, nuclear power and genetically modified organisms. This has created a well-established division of labour in which science-based regulation, framed as accountability or liability, determines the limits or boundaries of innovation, and where the articulation of socially desirable objectives—or what Rene von Schomberg describes as the 'right impacts' of science and innovation—is delegated to the market (von Schomberg, 2013). For example, with genetically modified foods, the regulatory framework is concerned with an assessment of potential risks to human health and the environment rather than with whether this is the model of agriculture we collectively desire.

This consequentialist and risk-based framing of responsibility is limited, because the past and present do not provide a reasonable guide to the future, and because such a framework has little to offer to the social shaping of science towards socially desired futures (Adam & Groves, 2011; Grinbaum & Groves, 2013). With innovation, we face a dilemma of control (Collingridge, 1980), in that we lack the evidence on which to govern technologies before pathologies of path dependency, technological lock-in, 'entrenchment' and closure set in. Dissatisfaction with a governance framework dependent on risk-based regulation and with the market as the core mediator has moved attention away from accountability, liability and evidence towards more *future*-oriented dimensions of responsibility—encapsulated by concepts of care and responsiveness—that offer greater potential for reflection on uncertainties, purposes and values and for the co-creation of responsible futures.

Such a move is challenging for at least three reasons: first, because there exist few rules or guidelines to define how science and technology should be governed in relation to forward-looking and socially desirable objectives (see Hajer, 2003, on the concept of the institutional void); second, because the (positive and negative) implications of science and technology are commonly a product of complex and coupled systems of innovation that can rarely be attributed to the characteristics of individual scientists (see Beck, 1992, on the concept of 'organised irresponsibility'); and third, because of a still-pervasive division of labour in which scientists are held responsible for the integrity of scientific knowledge and in which society is held responsible for future impacts (Douglas, 2003).

It is this broad context that guided our attempt to develop a framework of responsible innovation for the UK research councils (Owen et al., 2012; Stilgoe et al., 2013). Building on insights and an emerging literature largely drawn from STS, we started by offering a broad definition of responsible innovation, derived from the prospective notion of responsibility described above:

> Responsible innovation means taking care of the future through collective stewardship of science and innovation in the present. (Stilgoe et al., 2013: 1570)

Our framework originates from a set of questions that public groups typically ask of scientists, or would like to see scientists ask of themselves. Based on a meta-analysis of cross-cutting public concerns articulated in UK Sciencewise government-sponsored public dialogues on science and technology, we identified five broad thematic concerns that structured public responses: these were concerns with the purposes of emerging technology, with the trustworthiness of those involved, with whether people feel a sense of inclusion and agency, with the speed and direction of innovation, and with equity: i.e., whether it would produce fair distribution of social benefit (Macnaghten & Chilvers, 2014). This typology, which appears to be broadly reflective of public concerns across a decade or so of research and across diverse domains of emerging technology (amongst our own, see Grove-White et al., 1997; Macnaghten, 2004; Macnaghten & Szerszynski, 2013; Macnaghten et al., 2015; Williams et al., 2017), can be seen as a general approximation of the factors that mediate concern and that surface in fairly predictable ways when people discuss the social and ethical aspects of an emerging technology. If we take these questions to represent aspects of societal concern in research and innovation, responsible innovation can be seen as a way of embedding deliberation on these within the innovation process. From this typology we derived four dimensions of responsible innovation—anticipation, inclusion, reflexivity, and responsiveness (the AIRR framework)—that provide a framework for raising, discussing and responding to such questions. The dimensions are important characteristics of a more responsible vision of innovation, which can, we argue, be heuristically helpful for decision making on how to shape science and technology in line with societal values.

Anticipation is our first dimension. Anticipation prompts researchers and organisations to develop capacities to ask 'what if…?' questions, to consider contingency, what is known, what is likely, what are possible and plausible impacts. Inclusion is the second dimension, associated with the historical decline in the authority of expert, top-down policy making, and also the deliberative inclusion of new voices in the governance of science and technology. Reflexivity is the third dimension defined, at the level of institutional practice, as holding a mirror up to one's own activities, commitments and assumptions, being aware of the limits of knowledge and being mindful that a particular framing of an issue may not be universally held. Responsiveness is the fourth dimension, requiring science policy institutions to develop capacities to focus questioning on the three dimensions listed above and to change shape or direction in response to them. This demands openness and leadership within policy cultures of science and innovation such that social agency in technological decision-making is empowered.

To summarise, our framework for responsible innovation starts with a prospective model of responsibility, works through four dimensions, and makes explicit the need to connect with cultures and practices of science and innovation. Since its inception our framework is being put to use by researchers, research funders and research organisations alike. Indeed, since we developed the framework in 2012, one of the UK research councils, the Engineering and Physical Science Research Council (EPSRC) has made an explicit policy commitment to it (EPSRC, 2013; see also Owen, 2014). Starting in 2013, using the alternative 'anticipate–

reflect–engage–act' (AREA) formulation (see Murphy et al., 2016), the EPSRC has developed policies that set out its commitments to develop and promote responsible innovation, and its expectations both of the researchers it funds and of their research organisations.

Discussion and Conclusion

In this paper I have discussed four paradigmatic ways of governing science and technology. I began with the linear model in which science is represented as the motor of prosperity and social progress and in which the social contract for science is configured as that of the state and industry, providing funds for science in exchange for reliable knowledge and assurances of self-governed integrity. I then explored the dynamics and features which contributed towards a new social contract for science in which the organisation and governance of science became explicitly oriented towards the avoidance of harms and the meeting of predefined societal goals and so-called grand challenges. A co-production model of science and society was subsequently introduced as a more adequate understanding of how science and social order are mutually constitutive of each other, and of the implications of such an approach for science and democratic governance. Finally, I set out a framework of responsible (research and) innovation as an integrated model of aligning science with and for society.

These four models should not be seen as wholly distinct or unrelated. Typically, they operate in concert, sometimes harmoniously, other times less so, in any governance process. Nevertheless, the broad move beyond the linear model of science and society must be applauded, both because science devoid of societal shaping is clearly poorly equipped to respond to the societal challenges we collectively face, and also because the premises that underpin the linear model, such as the fact–value distinction, are clearly poorly aligned with contemporary intellectual debate. Unshackled from outdated distinctions, a framework of responsible research and innovation offers opportunities, tools and possibilities to make science and its governance more responsive to the question as to 'what kind of society do we want to be' (Finkel, 2018: 1).

References

Adam, B., & Groves, G. (2011). Futures tended: Care and future-oriented responsibility. *Bulletin of Science, Technology & Society, 31*, 17–27.

Beck, U. (1992). *The risk society. Towards a new modernity*. Sage.

Brooks, S. et al., (2009). *Silver bullets, grand challenges and the new philanthropy*. STEPS Working Paper 24, STEPS Centre, Brighton. Downloaded from http://www.ids.ac.uk/files/dmfile/STEPSWorkingPaper24.pdf (26.09.2018).

Bush, V. (1945). *Science – The endless frontier. A report to the President*. United States Government Printing Office.

Calvert, J. (2013). Systems biology: Big science and grand challenges. *BioSocieties, 8*, 466–479.

Chilvers, J., & Kearnes, M. (Eds.). (2016). *Remaking participation: Science, environment and emergent publics*. Routledge.

Collingridge, D. (1980). *The social control of technology*. Open University Press.

Douglas, H. (2003). The moral responsibilities of scientists (tensions between autonomy and responsibility). *American Philosophical Quarterly, 40*, 59–68.

Edgerton, D. (2004). The linear model did not exist. In K. Grandin, N. Worms, & S. Widmalm (Eds.), *The science-industry nexus: History, policy, implications* (pp. 31–57). Science History Publications.

EPSRC. (2013). *Framework for responsible innovation*. Downloaded from https://epsrc.ukri.org/index.cfm/research/framework/ (26.09.2018)

Etzkowitz, H., & Leydesdorff, L. (2000). The dynamics of innovation: From National Systems and 'Mode 2' to a triple helix of university-industry-government relations. *Research Policy, 29*, 109–123.

European Commission. (2007). *The European research Area: New perspectives*. Green Paper 04.04.2007. Text with EEA relevance, COM161, EUR 22840 EN. Office for Official Publications of the European Communities, Luxembourg.

European Commission. (2013). *Fact sheet: Science with and for Society in Horizon 2020*. Downloaded from: https://ec.europa.eu/programmes/horizon2020/sites/horizon2020/files/FactSheet_Science_with_and_for_Society.pdf (26.09.2018).

Finkel, A. (2018). *What kind of society do we want to be? Keynote address by Australian government chief scientist, human rights Commission 'human rights and technology' conference, four seasons hotel, Sydney*. Downloaded from: https://www.chiefscientist.gov.au/wp-content/uploads/Human-Rights-and-Technology.pdf (26.09.2018).

Flink, T., & Kaldewey, D. (2018). The new production of legitimacy: STI policy discourses beyond the contract metaphor. *Research Policy, 47*, 14–22.

Funtowicz, S., & Ravetz, J. (1993). Science for the post-Normal age. *Futures, 25*, 739–755.

Gibbons, M., Limoges, C., Nowotny, H., Schwartzman, S., Scott, P., & Trow, M. (1994). *The new production of knowledge: The dynamics of science and research in contemporary societies*. Sage.

Godin, B. (2006). The linear model of innovation: The historical construction of an analytical framework. *Science, Technology & Human Values, 31*, 639–667.

Grinbaum, A., & Groves, C. (2013). What is 'responsible' about responsible innovation? Understanding the ethical issues. In R. Owen, J. Bessant, & M. Heintz (Eds.), *Responsible innovation: Managing the responsible emergence of science and innovation in society* (pp. 119–142). Wiley.

Grove-White, R., Macnaghten, P., Mayer, S., & Wynne, B. (1997). *Uncertain world: Genetically modified organisms, Food and Public Attitudes in Britain*. Centre for the Study of Environmental Change.

Hajer, M. (2003). Policy without polity? Policy analysis and the institutional void. *Policy Sciences, 36*, 175–195.

Hessels, L., & van Lente, H. (2008). Re-thinking new knowledge production: A literature review and a research agenda. *Research Policy, 37*, 740–760.

Irvine, J., & Martin, B. (1984). *Foresight in science: Picking the winners*. Pinter.

Irwin, A. (2006). The politics of talk: Coming to terms with the 'new' scientific governance. *Social Studies of Science, 36*, 299–330.

Jasanoff, S. (1990). *The fifth branch: Science advisers as policymakers*. Harvard University Press.

Jasanoff, S. (2003). Technologies of Humility: Citizen participation in governing science. *Minerva, 41*, 223–244.

Jasanoff, S. (Ed.). (2004). *States of knowledge: The co-production of science and the social order*. Routledge.

Jasanoff, S. (2016). *The ethics of invention: Technology and the human future*. W. W. Norton Co.

Jasanoff, S., & Simmet, H. (2017). No funeral bells: Public reason in a 'post-truth' world. *Social Studies of Science, 47*, 751–770.

Jump, P. (2014, April 17). 'No regrets', says outgoing EPSRC chief David Delpy: 'Thick skin' helped research Council boss take the flak for controversial shaping capability measures. *Times Higher Education*. Downloaded from: https://www.timeshighereducation.com/news/no-regrets-says-outgoing-epsrc-chief-david-delpy/2012694.article (26.09.2018).

Ludwig, D., Pols, A., & Macnaghten, P. (2018). Organisational review and outlooks: Wageningen University and Research. In van der Molen, F., Consoli, L., Ludwig, D., Pols, A., Macnaghten, P. (Eds.) *Report from National Case Study: The Netherlands*. Deliverable 9.1, Responsible Research and Innovation Project, (pp. 29–54). Downloaded from: https://www.rri-practice.eu/wp-content/uploads/2018/09/RRI-Practice_National_Case_Study_Report_ NETHERLANDS. Pdf (26.09.2018).

Macnaghten, P. (2004). Animals in their nature: A case study of public attitudes on animals, genetic modification and 'nature'. *Sociology, 38*, 533–551.

Macnaghten, P. (2010). Researching Technoscientific concerns in the making: Narrative structures, public responses and emerging nanotechnologies. *Environment & Planning A, 41*, 23–37.

Macnaghten, P., & Chilvers, J. (2014). The future of science governance: Publics, policies, practices. *Environment & Planning C: Government and Policy, 32*, 530–548.

Macnaghten, P., & Szerszynski, B. (2013). Living the global social experiment: An analysis of public discourse on geoengineering and its implications for governance. *Global Environmental Change, 23*, 465–474.

Macnaghten, P., & Urry, J. (1998). *Contested natures*. Sage.

Macnaghten, P., Davies, S., & Kearnes, M. (2015). Online. Understanding public responses to emerging technologies: A narrative approach. *Journal of Environmental Planning and Policy*. http://goo.gl/7mOfDv

Merton, R. (1973). The normative structure of science. In N. Storer (Ed.), *The sociology of science* (pp. 267–278). University of Chicago Press.

Miller, C. (2004). Climate science and the making of a global social order. In S. Jasanoff (Ed.), *States of knowledge: The co-production of science and the social order* (pp. 247–285). Routledge.

Murphy, J., Parry, S., & Walls, J. (2016). The EPSRC's policy of responsible innovation from a trading zones perspective. *Minerva, 54*, 151–174.

National Council on Bioethics. (2012). *Emerging biotechnologies: Technology, Choice and the Public Good*. Nuffield Council on Bioethics.

National Research Council. (1983). *Risk assessment in the Federal Government: Managing the process*. National Academies Press.

Nowotny, H., Scott, P., & Gibbons, M. (2001). *Re-thinking science*. Knowledge and the Public in an Age of Uncertainty.

Owen, R. (2014). The UK Engineering and Physical Sciences Research Council's commitment to a framework for responsible innovation. *Journal of Responsible Innovation, 1*, 113–117.

Owen, R., Macnaghten, P., & Stilgoe, J. (2012). Responsible research and innovation: From science in society to science for society, with society. *Science and Public Policy, 39*, 751–760.

Owens, S. (2015). *Knowledge, policy, and expertise: The UK Royal Commission on environmental pollution 1970–2011*. Oxford University Press.

Perrow, C. (1984). *Normal accidents: Living with high-risk technologies*. Princeton University Press.

Polanyi, M. (1962). The republic of science: Its political and economic theory. *Minerva, 1*, 54–73.

Rip, A. (2016). The clothes of the emperor. An essay on RRI in and around Brussels. *Journal of Responsible Innovation, 3*, 290–304.

Rogers, E. M. (1962). *Diffusion of innovations*. Free Press of Glencoe.

Rose, N. (2006). *The politics of life itself: Biomedicine, power, and subjectivity in the twenty-first century*. Princeton University Press.

Stilgoe, J., Owen, R., & Macnaghten, P. (2013). Developing a framework of responsible innovation. *Research Policy, 42*, 1568–1580.

Stirling, A. (2008). 'Opening up' and 'closing down': Power, participation, and pluralism in the
social appraisal of technology. *Science, Technology and Human Values, 23*, 262–294.

Stirling, A. (2014). *Emancipating transformations: From controlling 'the transition' to culturing
plural radical progress*. STEPS Working Paper 64, STEPS Centre, Brighton.

von Schomberg, R. (2013). A vision of responsible research and innovation. In R. Owen, J. Bessant,
& M. Heintz (Eds.), *Responsible innovation: Managing the responsible emergence of science
and innovation in society* (pp. 51–74). Wiley.

Williams, L., Macnaghten, P., Davies, R., & Curtis, S. (2017). Framing fracking: Exploring public
responses to hydraulic fracturing in the UK. *Public Understanding of Science, 26*, 89–104.

Zwart, H., Landeweerd, L., & van Rooij, A. (2014). Adapt or perish? Assessing the recent shift
in the European research funding arena from 'ELSA' to 'RRI'. *Life Sciences, Society and
Policy, 10*, 11.

Chapter 6
Research and Use of Nuclear Energy—Its Ambivalence(s) in Historical Context

Horst Kant

Abstract The discovery of a scientific fact is initially not ambivalent; only in its possibilities of use does its ambivalence become apparent. But the socially evaluated positive uses and negative misuses of scientific knowledge cannot be clearly distinguished, and both can also appear ambivalent in turn. In this sense, consequently, scientists already bear responsibility for their discoveries. Moreover, in the case of experimental research, the question must be asked, to what extent its realization already interferes with what is happening and thus acquires an ambivalent character (which, however, must be distinguished from the ambivalence of its use). With those issues in mind, the article discusses the questions raised by the discovery and early use of nuclear energy.

Throughout the history of science, it has been repeatedly discussed and stated that scientific knowledge may be used for both good and evil (metaphorically: blessing and/or curse, etc.) for humanity, whose organized cognitive activity it represents, and that the perception and analysis of this phenomenon has consequences both for the individual and collective behavior of scientists and for the social regulation of scientific activity at the cognitive as well the applicative levels.

The approach to this problem was essentially on two levels. On the *first*—relatively superficial—level, science itself appears to be neutral (neither good nor bad),

Revised and modified version of a lecture at the conference "Ambivalence of Science" of the Gesellschaft für Wissenschaftsforschung on March 31, 2017 in Berlin. A variant of this article appeared under the title: Die Entdeckung der Kernenergie – Fluch oder Segen? Einige wissenschaftshistorische Betrachtungen. In: Radiochemie, Fleiß und Intuition – Neue Forschungen zu Otto Hahn. Ed. by Vera Keiser. Berlin, Diepholz: GNT-Verlag 2018, pp. 395–432; I would like to thank Hubert Laitko for reviewing the manuscript and for his suggestions, which are reflected, above all, in the introductory chapter of this article.

H. Kant (✉)
Max Planck Institute for the History of Science, Berlin, Germany
e-mail: kant@mpiwg-berlin.mpg.de

© The Author(s) 2022
H. A. Mieg (ed.), *The Responsibility of Science*, Studies in History and Philosophy of Science 57, https://doi.org/10.1007/978-3-030-91597-1_6

while its applications can be qualified as either good (use) or bad (misuse). It is assumed that the use (for the good) and misuse of scientific knowledge can be clearly distinguished from each other and that the problem can in principle be solved if the misuse is reliably prevented—whether by legal regulations, rational decisions of conscience by scientists or the interaction of both.

On the *second*—deeper—level, it becomes clear that the distinction between good and evil in relation to science is only feasible up to a certain limit, which depends on and changes with historical circumstances, but not absolutely. In its core zone, scientific recognition is always *both* an enabler of and a threat to human and social existence and evolution, because—not unlike the human work from which it has grown—it inevitably intervenes in the integrity of nature (and society). One cannot have one without the other, but must accept and deal with the contradictory unity of these two polar provisions. The ambivalence concept has its place on this level.[1] It remains weakly defined and controversial. The intention of this essay is not to intervene directly in these controversies with definitional efforts. Rather, the aim is to enrich the material basis for relevant efforts in the theory of science by presenting and discussing some "features" from the recent history of science. Whether or not the ambivalence concept may be more precisely defined or concretized for specific situations—what is decisive is not the duality of the two polar aspects alone, but their interrelation, their contradictory unity.

When we speak of good/bad or benefit/damage in everyday language, it is quite clear that we are dealing here with *evaluations*, references to human interests, or even to existential determinations of human existence. This fundamental reference is often obscured when the term "ambivalence" comes into play, which should probably be understood more as an expression of the attempt to reflect theoretically on an intuitively perceived and everyday-language problem of science and its social existence. If one assumes that the ambivalence concept in scientific analysis does not replace, but rather explicates, everyday language term pairs such as good/evil or benefit/harm, then one must at least demand that such an explication relates the *cognitive* and *ethical* aspects of science. If this does not happen, then the reflection is under-complex, and there is no reason to use the term "ambivalence" here. But if one establishes such a relationship, then extremely complex relationships arise.

The history of nuclear energy research is, as will be shown below, a prime example of such complications. A widespread but naive way of dealing with the ambivalence problem in nuclear physics is, for example, to claim that the use of nuclear physics knowledge for weapons development is evil, but its use for energy

---

[1] The term ambivalence was first introduced at the beginning of the 20th century by the Swiss psychiatrist Eugen Bleuler (1857–1939) [Bleuler, E.: Die Ambivalenz. Zürich: Schulthess & Co., 1914]. Actually, it simply means ambiguity or polyvalence, especially in psychology the coexistence of opposing feelings and thoughts. In the meantime, the term has found its way into colloquial language and is used in a wide variety of fields, especially by science and technology, and—as is so often the case—this term is usually interpreted only one-sidedly and is used primarily to indicate negative consequences of technical developments. Cf. e.g., Fratzscher, W.: Zu Risiken und Nebenwirkungen lesen Sie den Sicherheitsbericht oder fragen Sie... In: Sitzungsberichte der Leibniz-Sozietät 112 (2011), pp. 131–141 (here p. 131 f).

production is good. In the historical–political reality, however, the creation and further development of nuclear weapons took place in confrontational situations between hostile states or blocs with the consequence of an extreme relativization of the good–evil polarity: one's own weapons are good, whereas the (similar) weapons of the enemy are evil. In the use of nuclear physics for energy production, the ambivalence of cognition and action becomes apparent in a different but no less drastic way: (positively evaluated) energy production is inextricably linked to non-eliminable side- or consequential effects, which in some countries, including Germany, are considered so negative that they justify the phasing out of nuclear energy. Other countries, on the other hand, do not share this assessment and continue to develop and build new generations of nuclear power plants.

In a conversation in 1941 with Friedrich Houtermans (1903–1966) about the possibility of an atomic bomb, Max von Laue (1879–1960) is credited with the statement: "[…] an invention you don't want to make, you don't make either."[2] Here, Laue spoke wisely of invention and not of discovery. This is because the issue of impact assessment is much more complicated and complex for basic research than for applied research and technological developments, but this does not mean that the latter is "simpler." Nevertheless, Meyer-Abich stated unequivocally and correctly with regard to the year 1945:

> The atomic bomb was a direct result of basic research. So *there is no basic research* in the sense of a space free of responsibility, but whoever contributed to the discovery of nuclear fission is jointly responsible for the deaths of Hiroshima and Nagasaki. Otto Hahn and the others involved knew this and suffered under the burden of this responsibility.[3]

Carl Friedrich von Weizsäcker (1912–2007) made this observation already in 1970 and realized:

> Science cannot afford not to consider the effects it exerts on life under the motto that it seeks the truth and nothing else. Personally, I have never found it understandable that scientists have been of the opinion that if what science produces in technology is used by politicians or the military in such a way that scientists are unhappy, to say that science has been misused here.
>
> After all, science has provided these means, and it is of course responsible for the means it puts into other hands. If it delivers into a political structure which is not adequate to these means, means which are ominous in this structure, the least that can be asked of science is that it should think about how to change the structure which apparently cannot avoid producing these ominous effects. In this sense, then, self-reflection of science is a demand on science.[4]

So far, so good. But how can this work in daily practice?

---

[2] Quoted from Hoffmann, K.: Schuld und Verantwortung: Otto Hahn. Konflikte eines Wissenschaftlers. Berlin etc.: Springer 1993, p. 169.

[3] Meyer-Abich, K. M.: Die Idee der Universität im öffentlichen Interesse. In: M. Eigen et al: Die Idee der Universität: Versuch einer Standortbestimmung. Berlin etc.: Springer 1988, pp. 23–39 (here p. 25 f). (translation by the author)

[4] von Weizsäcker, C. F.: Die Macht der öffentlichen Meinung im Kampf gegen Einzelinteressen. In: Süddeutsche Zeitung, No. 166/1970 of 13. 7. p. 7. (translation by the author)

The different disciplines are differently sensitive to discourses on the ambivalence problem, and each one is so to a different degree at different stages of its development. Nuclear physics is a field in which these discourses have become spectacularly explosive in the twentieth century and, far beyond the experts, affect the general public. They are still current. The approach of this essay is not *systematic*, but *historical,* in view of the ambivalence problem. It traces the history of this area since the emergence of ideas about the atomic nucleus–shell structure, marks the places where relevant discourses arose, and sheds light on them more closely in their historical context. Another question is to what extent the conceptual tool of the ambivalence concept is already sufficient to analyze the phenomena, here presented from the perspective of the history of science, from the perspective of the theory of science.

***

Since the beginning of the twentieth century, scientists have been studying the structure of atoms and the properties of their components in greater detail;[5] until then, the question of the existence of atoms was still controversial among scientists. The first atomic models were created around 1910, including Joseph J. Thomson (1856–1940) with his raisin cake or plum pudding model (1904), Ernest Rutherford (1871–1937) in 1911 with his model of a positive (not yet further structured) nucleus and negative electrons orbiting around it, followed by Niels Bohr (1885–1962) with the further development of Rutherford's model on a quantum theoretical basis into a kind of planetary model, which in turn was refined by Arnold Sommerfeld (1868–1951) in 1915/16. The electron had already been introduced as a concept around 1874 and was first experimentally verified by J. J. Thomson in 1897;[6] there were still no plausible ideas about a possible structuring of the atomic nucleus. The Bohr–Sommerfeld atomic model, as we know it today (with a structured nucleus), only came into existence in the 1930s, and although it is only of historical significance today, it remains a vivid illustration in everyday life.

Finally, in 1919, Rutherford brought about the first artificial atomic transmutation by bombarding nitrogen atoms with alpha particles, observing that the nitrogen was transformed into oxygen, releasing a hydrogen nucleus; for this positively

---

[5]The presentation of the scientific–historical contexts partly follows the author's paper "Die Entdeckung der nuklearen Energie – Einige wissenschaftshistorische Betrachtungen." In: Technik & Technologie techne cum episteme et commune bonum (= Sitzungsberichte der Leibniz-Sozietät der Wissenschaften, Vol. 131). Edited by L.-G. Fleischer & B. Meier. Berlin: trafo Wissenschaftsverlag 2017, pp. 189–207. I would like to refer to two further short articles on this history that are worth reading: Szilard, L.: Creative Intelligence and Society: The Case of Atomic Research, The Background in Fundamental Science. In: The Collected Works Vol. 1, Scientific Papers; The MIT Press 1972, pp. 178–189; Stamm-Kuhlmann, Th.: Die Internationale der Atomforscher und der Weg zur Kettenreaktion 1874–1942. In: Salewski, M. (Ed.): Das nukleare Jahrhundert. Stuttgart: Franz Steiner 1998, pp. 23–40.

[6]Among others by Wilhelm Weber (1804–1891), Hermann von Helmholtz (1821–1894) and George Johnstone Stoney (1826–1911); the latter also proposed the term *electron* in 1891.

charged hydrogen nucleus he introduced the term *proton* in 1920. While the previously known natural nuclear transformations through alpha and beta decay led to elements in which the atomic weight decreased or remained the same, a nuclear transformation was now realized in which an increase in atomic weight occurred (i.e., an element was created that was higher in the periodic table than the original element).[7] For nuclear transformation experiments it now became necessary to use high-energy elementary particles, especially protons and electrons (Rutherford had still used natural alpha emitters for his experiments), and so Rutherford stimulated the development of electrical acceleration facilities, which succeeded in the early 1930s.

The year 1932 is often referred to as the *annus mirabilis* of nuclear physics: deuterium and the neutron and positron were discovered, which independently inspired Werner Heisenberg (1901–1976) in Leipzig and Dmitri Dmitrievich Ivanenko (1904–1994) in Moscow to develop a proton–neutron concept of the atomic nucleus, and in the USA the cyclotron was developed as a particle accelerator by Ernest O. Lawrence (1901–1958) and collaborators. Also in 1932, John D. Cockcroft (1897–1967) and Ernest T. S. Walton (1903–1995) succeeded with a slightly different accelerator design at Rutherford's laboratory in Cambridge, England, in achieving the first nuclear transformation using artificially accelerated particles: Lithium nuclei were converted into two helium nuclei by bombarding them with accelerated protons. This was at the same time an experimental confirmation of Albert Einstein's (1879–1955) equivalence relationship of mass and energy, formulated in 1905 as part of his special theory of relativity. Einstein derived this relationship for electromagnetic radiation, but concluded that it must also apply to all other forms of energy turnover. However, this Einsteinian relationship did not yet play a role in the first considerations about atomic energy. Only in the 1920s did it become relevant in atomic physics considerations (mass defect) and could then be experimentally confirmed in 1932 in the light of the new discoveries.

With these first transmutations of atomic nuclei, a new variant of the alchemists' old dream of producing gold through transmutation of elements came into focus. However, it must be clearly stated that physicists (and chemists) at that time were actually neither looking for this special possibility nor for a practically usable source of energy based on transmutation of atomic nuclei, but that it was primarily a matter of grasping the components and structure of atoms and their nuclei and understanding the mechanisms of such nuclear transmutations—i.e., in the truest sense of the claim, by Goethe's Faust, to "…detect the inmost force which binds the world, and guides its course."[8] The energetic relationships, insofar as they are important for the cohesion of the various components, were also discussed. However, the question of whether this energy could be used practically was still of secondary importance. Although it was recognized that enormous amounts of energy are con-

---

[7] In 1925 Rutherford's student Patrick S. Blackett (1897–1974) was able to visualize this process with the help of a Wilson cloud chamber and thus verified it experimentally.

[8] Goethe, Faust I (1808) lines 382–383; english translation by Bayard Taylor (1912) [The Project Gutenberg EBook: https://gutenberg.org/files/14591/14591-h/14591-h.html]

verted in relation to the quantities of substances involved, several orders of magnitude greater than the reaction heat of chemical reactions, no thought was given to this fact—with a few exceptions, which we will discuss in a moment.

The term *atomic energy*[9] was coined by the meritorious grammar school teachers Julius Elster (1854–1920) and Hans Geitel (1855–1923). In a lecture to the *Verein für Naturwissenschaften* in Braunschweig titled "Bemühungen, die Energiequelle der Bequerelstrahlen zu finden"[10] they concluded in 1899 that:

> [...] one will rather have to derive the energy source from the atom of the element in question itself. The thought is not far away that the atom of a radioactive element [...] will change into a stable state by releasing energy.[11]

The prescience of this insight becomes clear when one considers that at that time the idea of an atomic structure of matter was only just beginning to gain acceptance. Indeed, the first thoughts on the use of nuclear energy occurred relatively early on, but such ideas appeared overly utopian and were therefore not really taken seriously by the majority of the research community.[12]

Rutherford's student Frederick Soddy (1877–1956) gave a series of public experimental lectures in 1908, titled "The Interpretation of Radium."[13] In these he concluded by suggesting in vague terms that radium or radioactivity could in future help to identify and control the original sources of energy—thereby anticipating radium as an inexhaustible source of energy. Many of the basic ideas in Soddy's book are absolutely correct, even if the atomic models were still very poor: Bohr's atomic model would not be proposed until 1913, and no one yet had any idea of the structure of atomic nuclei.

---

[9] Today, when it comes to energy production from the atom, one prefers to speak of nuclear energy, because in fact it is about the energetic relationships in the nucleus and not between the nucleus and the "atomic shell"; but often, both expressions are used synonymously.

[10] In 1896, Antoine Henri Becquerel (1852–1908) in Paris—shortly after the discovery of X-rays by Wilhelm Conrad Röntgen (1845–1923)—had discovered radioactive radiation, thus opening up a new field of research. Among the pioneers of radioactivity research were Marie (1867–1934) and Pierre (1859–1906) Curie, Ernest Rutherford (1871–1937), Frederick Soddy (1877–1956), Otto Hahn (1879–1968) and Hans Geiger (1882–1945).

[11] Elster, J. & Geitel, H.: Weitere Versuche an Becquerelstrahlen. Annalen der Physik, 305(9) N.F. 69(1899)1, pp. 83–90, here p. 88. See also Fricke, R.: J. Elster & H. Geitel – Jugendfreunde, Gymnasiallehrer, Wissenschaftler aus Passion. Braunschweig: Döring 1992, p. 116.

[12] Nevertheless, this realm of utopia flourished and was found to differ only slightly from reality. In 1910, for example, the American radium researcher Everard Hustler described both the possibility of a nuclear weapon (albeit on the basis of a completely different operating principle) and radium as a medical panacea for the expected "century of radium" [in: Brehmer, A. (Ed.): Die Welt in 100 Jahren. Berlin 1910, pp. 245–266 (new edition Hildesheim etc.: Olms 2012)]. For a discussion of these predictions see, among others, Steger, F. & Friedmann, H.: Radium – Ein faszinierendes Element: Segen oder Fluch? part 3: Radium in der Medizin, in Industrieprodukten und im privaten Bereich. In: Strahlenschutz aktuell 46(2011)1, pp. 7–47 (esp. pp. 44–45).

[13] Soddy, F.: The interpretation of radium. 1909. – German: Die Natur des Radiums. [1909] (= Ostwalds Klassiker der exakten Wissenschaften Bd. 289). Frankfurt am Main: Harri Deutsch 2002.

Soddy's description inspired the well-known British utopian writer H. G. Wells (1866–1946) to write his novel "The World Set Free," which was published in 1914 (in German only in 1985). The book begins with the statement: "The history of mankind is the history of the attainment of external power."[14] A few pages later, it states: "The latent energy of coal and the power of steam waited long on the verge of discovery, before they began to influence human lives."[15] Wells understood human history here as energy history. In Wells' novel, the professor character[16] states:

> Radium is an element that is breaking up and flying to pieces. [...] And we know now that the atom, that once we thought hard and impenetrable [...], is really a reservoir of immense energy. [...]

Wells then lets his lecturer show a bottle of uranium oxide and continues:

> And in this bottle [...] there slumbers at least as much energy as we could get by burning a hundred and sixty tons of coal. [...] But at present no man knows, [...] how this little lump of stuff can be made to hasten the release of its store. [...].[17]

The novel mainly deals with the consequences of unleashing the newly discovered nuclear energy, both in its civil applications and military use as atomic bombs. In Wells' novel the decisive discovery that enabled this was made by a chemist in 1935![18] The English edition contains a dedication (unfortunately omitted from the German version): "To Frederick Soddy's 'Interpretation of Radium' [...]."[19]

After Soddy, Francis William Aston (1877–1945) was one of the few scientists who took the possibility of using atomic energy seriously.[20] As early as 1919 he postulated the extremely high-energy fusion of hydrogen into helium. In 1922, Aston was awarded the Nobel Prize in Chemistry; and he concluded his acceptance lecture by pointing out that the fusion of hydrogen atoms could be the energy source for the Sun, and finished by remarking:

---

[14] Here cited after the English Ebook edition: Wells: The world set free. Project Gutenberg 2006 [http://www.gutenberg.org/ebooks/1059] (Accessed: 28 February 2017). – In German: Wells, Herbert G.: Befreite Welt. Wien/Hamburg: Paul Zsolnay 1985, p. 7.

[15] Ibid, section 4.

[16] Soddy corresponds to the character Rufus in Wells' novel.

[17] Ibid, section 8.

[18] This is surprisingly close to the actual year of discovery.

[19] The complete dedication reads: "This Story, which owes long passages to the eleventh chapter of that book, acknowledges and inscribes itself." [http://www.gutenberg.org/ebooks/1059] (Accessed: 28 February 2017).

[20] We should also mention the German physicochemist Walther Nernst (1864–1941), Nobel Prize winner in 1920, who in 1912 at the *Versammlung Deutscher Naturforscher und Ärzte* in Münster i.W. remarked in his lecture "Zur neueren Entwicklung der Thermodynamik": "The discovery of the radioactive decay of the elements has introduced us to sources of energy of a power of which we had no conception before." (Nernst, Walther: Das Weltgebäude im Licht der neueren Forschung. Berlin: Springer 1921, p. 2, translated). In 1921, however, he added: "However, one should beware of the illusion that the technical extraction of the energy quantities available here has come within reach." (Ibid, p. 23, translated).

> Should the research worker of the future discover some means of releasing this energy in a form which could be employed, the human race will have at its command powers beyond the dreams of scientific fiction; but the remote possibility must always be considered that the energy once liberated will be completely uncontrollable [...].[21]

But the reality was still completely different. Although, as mentioned above, Rutherford had first demonstrated nuclear transmutation in 1919, the process was far from releasing energy. On the contrary, the energy released in this process was actually less than the energy required by the alpha particle.

And so, it is not surprising that Rutherford said in a lecture at the *British Association for the Advancement of Science* meeting in Leicester in September 1933:

> These transformations of the atom are of extraordinary interest to scientists but we cannot control atomic energy to an extent which would be of any value commercially, and I believe we are not likely ever to be able to do so. A lot of nonsense has been talked about transmutation. Our interest in the matter is purely scientific, and the experiments which are being carried out will help us to a better understanding of the structure of matter.[22]

This was reported as the famous "moonshine" quotation: *Nature* wrote that Rutherford had said that anyone who sees atomic transformations as a source of energy talks "moonshine."[23] He confirmed this opinion again in 1936, although perhaps not quite so emphatically:

> [...] here seems to be little hope of gaining useful energy from the atoms by such methods. [...] At the moment, however, the natural radioactive bodies are the only known sources for gaining energy from atomic nuclei, but this is on far too small a scale to be useful for technical purposes.[24]

Other eminent scientists, such as Einstein and Bohr, were of the same opinion. Finally, the important Croatian-American inventor Nikola Tesla (1856–1943) should be mentioned here, certainly someone who was open to unusual ideas, who is also credited with stating, in 1931, that the idea of atomic energy is illusory and that it can be used for neither civilian nor military purposes.[25] Even the leading

---

[21] Aston, F. W.: Mass spectra and isotopes. Nobel lecture 1922. In: Nobel Lectures Chemistry 1922–1941. Amsterdam: Elsevier 1966, p. 20.

[22] Quoted from Eve, A. S.: Rutherford—Being the life and letters of the Rt. Hon. Lord Rutherford, O.M. The Macmillan Company New York & The University Press Cambridge/England 1939, p. 374. Rather jokingly, however, Rutherford had already in 1903 expressed the "disturbing idea" that with a suitable detonator a huge explosion wave could be started by atomic decay, which could turn the whole mass of the globe into helium (see Weart, Sp.: Nuclear Fear. A History of Images. Harvard University Press 1988, p. 18).

[23] [A.F.]: Atomic Transmutation. In: Nature 132(1933)3333, Sept. 16, pp. 432–433 (here p. 433). See also: Jenkin, J. G.: Atomic Energy is "Moonshine": What did Rutherford *Really* Mean? In: Physics in Perspective 13(2011)2, pp. 128–145.

[24] Rutherford, E.: The Transformation of Energy. In: Nature 137(1936, Jan 25)3456, S. 135–137 (here p. 137).

[25] Quoted after [http://www.nur-zitate.com/autor/Nikola_Tesla](Accessed: 6 October 2016). Analogous in: Tesla, 75, Predicts New Power Source. In: New York Times of 5 July 1931. Cf. also Cheney, M.: Nikola Tesla – Erfinder, Magier, Prophet. Aachen: Omega 2005, p. 258 f.

Soviet physicist Pyotr L. Kapiza (1894–1984), who had worked for several years in Rutherford's laboratory, was still convinced at the beginning of 1940—when nuclear energy had already become a reality—that the use of nuclear energy was not to be expected, and in doing so referred to Rutherford. Restrictively, he remarked that perhaps something else would be discovered, but that this seemed unlikely.[26]

However, a young Hungarian physicist, Leo Szilard (1898–1964), who first emigrated to Germany in 1921 then subsequently to England when the Nazis came to power, had just read H. G. Wells' novel; he also read the aforementioned statement by Rutherford in the *Times*, but did not consider the prospect of atomic energy at all unrealistic. Szilard then developed the idea of the atomic chain reaction (which Wells could not yet envision in his fiction) and was even secretly granted a patent through Britain's Royal Navy a year later.[27] However, even Szilard was unable to specify which elements might be suitable for such a process. Also, Ernest O. Lawrence (1901–1958), for example, the creator of the cyclotron, said: "I have no opinion as to whether it can ever be done, but we're going to keep on trying to do it."[28]

So much for the visions or fictions of nuclear energy, which certainly existed in the first decades of the twentieth century.[29] But most of the "serious" scientists did not really believe in the prospect of nuclear energy nor see any way of testing its feasibility. In science fiction literature, however, this possibility of energy production was still being considered, although mostly not as profoundly as in Wells' work.[30] So the question is, to what extent scientists should or should not have taken these fictions seriously? Szilard at least tried.

***

The latest results of nuclear research also inspired a young German physicist. Carl Friedrich von Weizsäcker had studied at the then physical centers of Germany: Berlin, Göttingen, and Leipzig. Weizsäcker received his doctorate from Leipzig in the summer of 1933 with a thesis on nuclear physics under Werner Heisenberg, who had been physics professor there since 1927. In 1936 Weizsäcker habilitated there with a thesis on nuclear forces. He then went to Berlin, to the newly founded Kaiser Wilhelm Institute for Physics, which was headed by the Dutch physicist Peter

---

[26] Cf. interview for the children's magazine "Detskaja Literatura" 1940, No. 4, pp. 18–23, excerpts printed in: Atomnoj Proekt SSSR, Vol. 1, Part I. Moskva: Izd. Nauka 1998, pp. 93–94.

[27] Szilard, L.: Creative Intelligence and Society: The Case of Atomic Research, The Background in Fundamental Science. In Szilard, L: The Collected Works Vol. 1, (loc. cit.)., pp. 178–189 (here p. 183).

[28] Lawrence expressed this as a comment on the report on Rutherford's statement in the *New York Harold Tribune* of 12.9.1933. (Quoted after Weiner, Ch.: Physics in the Great Depression. In: Physics Today 23 (October, 1970), pp. 31–38 (here p. 35)).

[29] See also Strub, E.: Soddy, Wells und die Atombombe. In: Physik Journal 4(2005)7, pp. 47–51.

[30] It is not possible to go into these aspects in greater depth here; reference is only made as an example from German-language literature to Hans Dominik (1872–1945) and his novel "Der Brand der Cheopspyramide" (1925/26).

Debye (1884–1966), who had also previously worked in Leipzig.[31] As the KWI for Physics was still under construction and would not start work until spring 1937, Weizsäcker took the opportunity in autumn 1936 to spend a quarter of a year at the Kaiser Wilhelm Institute of Chemistry, where he was to represent Max Delbrück (1906–1981), whom he knew from Copenhagen meetings at the Bohr Institute.[32] Delbrück was then a so-called "house theorist" for Lise Meitner (1878–1966) and Otto Hahn, the leading German researchers in the field of radioactivity, but was already more interested in biochemistry. At the end of 1936 Weizsäcker's monograph on nuclear physics "Die Atomkerne—Grundlagen und Anwendungen ihrer Theorie" [The atomic nuclei—basics and applications of their theory] appeared, which attempted to bring together the latest findings of recent years.[33] In his preface written in Berlin in September 1936, he outlined his concern:

> The rapidly growing body of experience about atomic nuclei does not yet permit the formulation of an exhaustive theory, but it does consistently show the correctness and fertility of certain basic theoretical ideas. This book addresses on the one hand readers with an experimental background who want to gain an overview of the secure part of this core theory, and on the other hand theoreticians who are interested in its further development.[34]

In his review of this book, Delbrück wrote that the study of atomic nuclei is currently the most popular field of work for physicists and stated that Weizsäcker's book is "[...] the first theoretical account of the subject" and offers "[...] a concise summary in clear, understandable language of our present knowledge of atomic nuclei and all the theoretical aspects which have proved useful in their analysis [...]."[35]

Remarkable for our consideration here is the illustration on p. 51 of Weizsäcker's book (see Fig. 6.1), because it shows (in today's terms) the magnitude of the nuclear binding energy as a function of the nuclear mass. It follows from the curve that useful energy can be obtained in the area of light nuclei by nuclear fusion, whereas in the area of heavy nuclei it can be obtained by nuclear fission (into two medium-heavy nuclei). However, physicists at that time did not seriously consider these possibilities, because other convictions stood in opposition, including the basic statement that nuclear transformations can only occur between elements that are

---

[31] Originally, the KWI for Physics was founded in 1917 under Einstein's leadership, but at that time without institutional buildings and laboratories. At the latest since the end of the 1920s the idea of an institute building was pursued. (Cf. inter alia Kant, H.: Max-Planck-Institut für Physik Berlin – München. In: Denkorte. Max-Planck-Gesellschaft und Kaiser-Wilhelm-Gesellschaft. Brüche und Kontinuitäten 1911–2011. Edited by P. Gruss & R. Rürup. Dresden: Sandstein 2010, pp. 316–323).

[32] Cf. inter alia Kant, H.: Vom KWI für Chemie zum KWI für Radioaktivität: Die Abteilung(en) Hahn/Meitner am Kaiser-Wilhelm-Institut für Chemie. In: Dahlemer Archivgespräche, vol. 8, Berlin 2002, pp. 57–92.

[33] von Weizsäcker, C. F.: Die Atomkerne – Grundlagen und Anwendungen ihrer Theorie. (= Physik und Chemie und ihre Anwendungen in Einzeldarstellungen. Vol. 11). Leipzig: Akademische Verlagsgesellschaft 1937.

[34] Ibid. (translation by the author)

[35] Delbrück, M.: Review of von Weizsäcker, C. F.: Die Atomkerne. In: Physikalische Zeitschrift 38(1937) p. 388. (translation by the author)

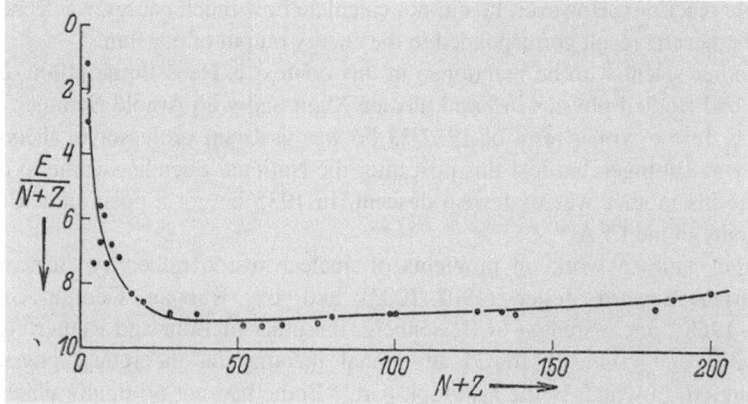

Fig. 6.1 Magnitude of the nuclear binding energy as a function of the nuclear mass. Weizsäcker referred in the ordinate to the packing fraction, which is the quotient of mass defect and nucleon number. It is a measure of the relative stability of atomic nuclei (N = number of neutrons, Z = number of protons; N + Z = nucleon number). (Source: von Weizsäcker, 1937, Die Atomkerne – Grundlagen und Anwendungen ihrer Theorie, p. 51)

close together in the periodic table of elements.[36] This "omission" can be observed from a historical perspective, but probably none of the scientists involved can be blamed for it.

Weizsäcker was very much interested in astronomy from his earliest youth. Since the common atomic models showed parallels to planetary systems, he also looked for connections there. And since nuclear physics was now "in," he thought that the energy source of the stars might be a nuclear reaction (Aston had already postulated this); that question was also "in" among astrophysicists at that time.[37] Weizsäcker published his thoughts on this in the journal *Physikalische Zeitschrift* and also presented them to the *Berlin Physical Society* [Physikalische Gesellschaft zu Berlin] in 1938.[38]

Astrophysicists assumed that the Sun mainly consisted of hydrogen and helium. The energy production mechanism therefore had to be based primarily on the nuclei of those elements. Weizsäcker suggested several possible basic reactions, and ultimately it was found that helium is produced from hydrogen—releasing large amounts of energy—by a fusion reaction. Weizsäcker also indicated various other

---

[36] In other words, their atomic numbers (equal to the number of protons in the nucleus) differ only by 1.

[37] At that time Weizsäcker had just read the book by the eminent British astrophysicist Sir Arthur Eddington (1882–1944) *The Internal Constitution of Stars* (published in 1926), and Eddington had made a similar conjecture.

[38] von Weizsäcker, C. F.: Über Elementumwandlungen im Innern der Sterne. In: Physikalische Zeitschrift 38(1937)6, pp. 176–191 & 39(1938)16, pp. 633–646.

possible reactions. However, he did not calculate how much energy was generated nor whether the result corresponded to the energy output of our Sun.[39]

Another scientist to be mentioned in this context is Hans Bethe (1906–2005). Bethe had studied physics in Frankfurt am Main and with Arnold Sommerfeld in Munich. In the winter term of 1932/33 he was assistant professor of theoretical physics at Tübingen but lost this post after the National Socialists came to power because his mother was of Jewish descent. In 1935 he got a position at Cornell University in the USA.[40]

Bethe came to work on problems of nuclear fusion rather by chance. The Hungarian Edward Teller (1908–2003) and the Russian George Gamow (1904–1968), one a student of Heisenberg, the other of Bohr and Rutherford and both also emigrants, organized an annual meeting on theoretical physics in Washington, in which Bethe also took part.[41] Bethe had not originally planned to participate in the conference of March 1938, since it was to deal with astrophysical questions, but Teller had persuaded him to do so because the aim of this 4th conference was to familiarize as many physicists as possible with current astrophysical problems, so that they might perhaps be able to contribute something to them.[42] Bethe remembered with the following words:

> At this conference the astrophysicists told some of us physicists what stars are about, how they are made, what distribution of density and pressure they have and so on, and then they ended up with the question where does the energy come from? Everybody of course agreed that the energy must come from nuclear reactions but what nuclear reactions? They were searching at the time for too much, viz. they were trying at the time to solve simultaneously the problem of the buildup of elements and the problem of production of energy in the stars. It was just the removal of this coupling that made it possible to solve the problem.[43]

First Bethe had worked out the simple fusion reaction between two hydrogen nuclei with Charles L. Critchfield (1910–1994), a student of Gamow and Teller, referring to Weizsäcker's article of 1937.[44] The result was quite consistent with the value given by the astrophysicists for the energy production of the Sun, but with larger stars this simple reaction was not possible. Bethe then calculated systematically through the periodic table and looked for nuclei that could react with hydrogen nuclei. Finally, he found what he was looking for in carbon. In this process, the carbon merely serves as a catalyst by first fusing with a hydrogen nucleus to form a nitrogen isotope, which then decays again into another carbon isotope, which in the

---

[39] There is insufficient space here to provide further details.

[40] Bethe received a professorship there in 1937, became a US citizen in 1941 and was involved in the Manhattan Project, including as head of the Department of Theoretical Physics there.

[41] Coincidentally, Bethe and Teller had been on the same ship on the passage to the USA. Cf. Hargittai, I.: Judging Edward Teller. Prometheus Books 2010, p. 103.

[42] Cf. inter alia Bethe, H. A.: Energy on Earth and in the Stars. In: From a Life of Physics. World Scientific 1989, pp. 1–18.

[43] Ibid., p. 11.

[44] Bethe, H. A. & Critchfield, C. L.: The Formation of Deuterons by Proton Combination. In: Physical Review 54(1938)4, pp. 248–254.

next step fuses again with a hydrogen nucleus and so on. Finally, the overall result of such a carbon–nitrogen cycle is the fusion of four hydrogen nuclei to form a helium nucleus (alpha particle), whose mass is about 1% less than the mass of the four hydrogen nuclei (protons). This mass defect is almost entirely converted into energy according to Einstein's equation $E = mc^2$. This carbon–nitrogen cycle is also known today as the Bethe–Weizsäcker cycle. It only occurs at temperatures in excess of 14 million Kelvin.

Further possible fusion processes are not discussed here. The high temperatures make it clear that it was completely illusory to think of a practical use at that time. That is why the physicists did not pursue the matter any further, because the astrophysicists' question was first solved for them in principle.

Bethe was awarded the Nobel Prize in Physics in 1967 for his work; Weizsäcker was left empty-handed because he had essentially only looked at the matter qualitatively, and since both had worked independently, Bethe had thus made the more complete achievement. Weizsäcker saw it that way himself.[45]

***

The discoveries made in the *annus mirabilis* 1932 prompted the young Italian physicist Enrico Fermi (1901–1954) to set his newly constituted research group in Rome an interesting task. Fermi had quickly realized that the neutron discovered by James Chadwick (1891–1974) should be better suited to bringing about atomic transformations because of its neutral charge than the positively charged alpha particle previously used for such purposes. The final impetus for his concrete investigations came with the discovery of artificial radioactivity in early 1934 by Irène (1897–1956) and Frédéric (1900–1958) Joliot-Curie.[46] In the following years Fermi's group showed that almost all elements can be transformed into radioactive elements by bombarding them with neutrons. This was based on the following model: when a low-energy (i.e., "slow") neutron collides with an atomic nucleus, it can be absorbed into the nucleus; in the newly formed nucleus, it is then transformed into a proton, emitting a beta particle, and the atomic number of the element increases by one unit.[47]

The observed transformation mechanism led, among other things, to the assumption that when bombarding uranium, the heaviest naturally occurring element—and thus the last element in the Periodic Table of Natural Elements with the atomic

[45] Cf. Schaaf, M.: Weizsäcker, Bethe und der Nobel Prize. In: Carl Friedrich von Weizsäcker: Physik – Philosophie – Friedensforschung. (= Acta Historica Leopoldina No. 63), edited by Klaus Hentschel and Dieter Hoffmann. Stuttgart: Wissenschaftliche Verlagsgesellschaft 2014, pp. 145–156.

[46] In Detail cf. inter alia De Gregorio, A.: Neutron physics in the early 1930s. In: Historical Studies in the Physical and Biological Sciences 35(2005)2, S. 293–340.

[47] Kant, H.: Von den falschen Transuranen zur Kernspaltung – die Atomphysiker Enrico Fermi und Lise Meitner. In: Italien und Europa. Der italienische Beitrag zur europäischen Kultur. Edited by Franziska Meier & Italien-Zentrum der Universität Innsbruck; Innsbruck: Studien Verlag 2007, pp. 171–186.

number 92—so-called transuranium elements would have to be created, i.e., "artificially produced" elements that would follow uranium in the Periodic Table.

Initial reactions to Fermi's papers pointed out, inter alia, that it could not be excluded that instead of the element 93 (as claimed by Fermi), element 91 could have been created. But element 91 (i.e., protactinium) was a domain of the Berlin researchers Otto Hahn and Lise Meitner, who had discovered it in 1918. So, at the end of 1934, they also turned to this research.

It was not least the political circumstances in Germany that forced Hahn and Meitner to resume closer direct cooperation again after 1934.[48] Meitner persuaded Hahn to carry out a joint review of the Fermi results; Fritz Straßmann (1902–1980), Hahn's assistant since 1929, was called in for the necessary chemical analyses. The assertion of the possible creation of protactinium was soon disproved, but the question remained as to what really occurs during neutron irradiation.[49]

Between 1934 and 1938, Hahn, Meitner, and Straßmann published 15 papers on questions of the artificial transmutation of uranium by neutrons. On the basis of the Fermi hypothesis, transuranium elements were expected as transmutation products. In 1937, the three scientists seemed to have found a solution and presented three possible decay series of uranium in two papers—one chemically and one physically oriented.[50] The reaction products resulting from the irradiation of uranium (and other elements) with neutrons first had to be separated chemically in order to then determine their radioactivity by physical means. Since the quantities involved were very small—almost "imponderable"—particularly difficult chemical analysis methods were necessary, but which Hahn and his staff in particular mastered excellently.

The nuclear transformations also produced radioactive products that could only be separated with barium and in which radium isotopes were therefore suspected;[51] this part of the research program, however, raised more questions than answers could be found. The classification difficulties were not least due to the fact that at that time the so-called actinide series in the periodic table of the elements was not yet known and therefore the classification of the last three natural elements was somewhat different than today.[52]

---

[48] Meitner was Austrian and came from a Jewish family, but had already converted to the Protestant faith in 1908.

[49] Kant, H.: Die radioaktive Forschung am Kaiser-Wilhelm-Institut für Chemie von den Anfängen bis zum deutschen Uranprojekt. In: Kant, H. & Reinhardt, C. (Ed.): 100 Jahre Kaiser-Wilhelm-/Max-Planck-Institut für Chemie (Otto Hahn Institut). Facetten seiner Geschichte. (= Veröffentlichungen aus dem Archiv der Max-Planck-Gesellschaft, Vol. 22) Berlin 2012, pp. 53–98.

[50] Hahn, O., Meitner, L. & Straßmann, F: Über die Transurane und ihr chemisches Verhalten. In: Berichte der Deutschen Chemischen Gesellschaft 70(1937) pp. 1374–1391; Meitner, L., Hahn, O. & Straßmann, F.: Über die Umwandlungsreihen des Urans, die durch Neutronenbestrahlung erzeugt werden. In: Zeitschrift für Physik 106(1937)3/4, pp. 249–270.

[51] Barium is a so-called homologue of radium, i.e., it is located in the 2nd main group below radium in the periodic table.

[52] At that time uranium was classified in the 6th subgroup of the 7th period; today it is found at the 3rd position in the actinide series.

That barium could indeed have been formed, as some analysis results suggested, was beyond any reasonable assumption at the time. It is interesting to refer in this context to an episode from the KWI for Chemistry: Straßmann reported that as early as 1936, between two routine measurements that he carried out as part of the research program, he also carried out a measurement with barium at night, which he found interesting. But Lise Meitner said the next morning: "Oh, you had better leave that to us physicists—you can toss this into the wastebasket."[53] And Straßmann stated: "What I threw away at that time was already the proof of the formation of barium from uranium after bombarding with slow neutrons—that is, nuclear fission."[54] Is this really an example of ambivalence in experimental research, as for instance Parthey argues,[55] when rejecting an experimental result that obviously contradicts the accepted theory about the process? After all, at this point in time at least the first foundations of an "alternative" theory already existed, which then confirmed this result. From my point of view this remains an open question.

Although the chemist Ida Noddack (1896–1978) had already expressed doubts about Fermi's transuranium results in 1934, and had just expressed the suspicion that they might also be fission products of uranium, she had no possibility at that time to verify this experimentally. And the physicists concluded, from the theories about the nucleus, that nuclear transformations into neighboring elements were possible, but not a "bursting" of the nucleus. When Ida Noddack's publication was taken up by physicists at the time, a memoir by Emilio Segrè (1905–1989) from the Fermi group shows how dubiously they regarded Noddack's publication:

> We did not seriously entertain the possibility of nuclear fission, although it had been mentioned by Ida Noddack, who sent us a reprint of her work. The reasons for our blindness, shared by Hahn and Meitner, the Joliot-Curies and everybody else working on the subject, is not clear to me even today.[56]

Soon after the publication of Hahn and Meitner's decay series, astonishing news came from Paris. Irène Joliot-Curie and her colleagues had found a reaction product with a slightly different method of analysis than was customary in Berlin, a product that the Berliners had probably overlooked until then; but in interpreting this product, the Parisians indulged in "wild speculations" that Meitner found theoretically untenable, and soon lost all interest in the reaction product, which she dubbed "Curiosum."[57] In mid-1938, she was forced to leave Germany as a result of the German annexation of Austria, and found asylum and a modest research opportunity in Sweden.

---

[53] Straßmann, F.: Kernspaltung. Privatdruck, Mainz 1978, p. 31. (translation by the author)

[54] Krafft, F.: In the Shadow of the Sensation. Life and Work of Fritz Straßmann. Weinheim etc.: Publisher Chemie 1981, p. 218. (translation by the author)

[55] Cf. Parthey, H.: Institutionalization, interdisciplinarity, and ambivalence in research situations. See Ch. 7 of this volume).

[56] Segrè, E.: A Mind Always in Motion. The Autobiography of Emilio Segrè. Berkeley & Oxford: University of California Press 1993, p. 91; Lise Meitner, too, was very critical of Noddack's considerations; cf. Krafft, Fritz: Im Schatten der Sensation. op. cit.

[57] Krafft, F.: Im Schatten der Sensation; op. cit., p. 79.

In the autumn of 1938 a new surprising publication came from Paris about possible new reaction products in the transuranium search, and Hahn and Straßmann decided, after initial reluctance, to resume their series of experiments. In late October 1938 Hahn wrote to Meitner that they had now actually found this Parisian "body." Several letters were exchanged between them. On Monday December 19, 1938 Hahn wrote the following famous lines to Meitner:

> [...] It is now just 11 p.m.; [...] Actually there is something about the "radium isotopes" that is so remarkable that for now we are telling only you. [...] We are coming steadily closer to the frightful conclusion that our Ra isotopes do not act like Ra, but like Ba [...] Perhaps you can come up with some sort of fantastic explanation. We know ourselves that it *can't* actually burst apart into Ba.[58]

This was now the famous actual discovery of uranium nuclear fission.

Hahn was well aware of the significance of this discovery—although certainly not yet all its consequences—as well as of the competitive situation in which he found himself, at least with the Paris research group. On December 23, 1938 he submitted the manuscript of the corresponding article to the journal *Die Naturwissenschaften* and, thanks to his good relations with the editor, this article was printed in the first issue of 1939, which appeared on Friday 6 January.[59] Hahn had also sent the manuscript to Meitner and it reached her on December 30, 1938.

Together with her visiting nephew Otto Robert Frisch (1904–1979), who—also an emigrant—worked at Bohr's institute in Copenhagen, Meitner then succeeded in those days around the turn of the year 1938/39, taking into account the droplet model[60] that had been known for several years and Bohr's theory of the so-called "compound nucleus" developed from it 3 years earlier, in achieving the "fantastic explanation" that Hahn had hoped for—in other words: nuclear fission could indeed be explained within the framework of the physical theories that were already available! Thus, the joint work of the 1930s came to a crowning conclusion after all—Hahn and Straßmann provided the irrefutable chemical–experimental results, Meitner and Frisch the physical–theoretical explanation. However, they had found something other than what they were originally looking for!

The corresponding work by Meitner and Frisch was sent to the English journal *Nature* on January 16, 1939.[61]

---

[58] Quoted after Krafft, F.: Im Schatten der Sensation; op. cit., p. 263 f.

[59] Hahn, O. & Straßmann, F.: Über den Nachweis und das Verhalten der bei der Bestrahlung des Urans mittels Neutronen entstehenden Erdalkalimetalle. In: Die Naturwissenschaften 27(1939, 6. Jan)1, S.11–15.

[60] First introduced by George Gamow, further developed by others. Weizsäcker's comments in his 1937 book are also based on this.

[61] Meitner, L. & Frisch, O. R.: Disintegration of Uranium by Neutrons: A New Type of Nuclear Reaction. In: Nature 143(1939, Feb. 11) 3615, pp. 239–240. Following his return to Copenhagen, Frisch had by then been able to demonstrate nuclear fission in a physical experiment: If one assumes the fact of fission and bases the corresponding calculations on it, the expected amount of energy released can be measured relatively easily. His corresponding article appeared a week after the joint essay with Meitner.

After the paper by Hahn and Straßmann had been published, Weizsäcker at the KWI for Physics had also soon found the correct theoretical explanation, as had Siegfried Flügge (1912–1997), since 1937 Delbrück's successor as the new "house theorist" at the KWI for Chemistry, together with Meitner's long-time assistant Gottfried von Droste (1908–1992), whose corresponding paper was submitted on January 22, 1939 (published at the beginning of March), i.e., only 1 week after Meitner and Frisch.[62] Since Hahn had not revealed anything about the discovery at the Institute—in order to give Meitner the chance to find the physical explanation and thus somehow bring together "the old team" in this discovery—Berlin colleagues were, understandably, somewhat "angry" about this "secrecy."[63]

It was also Flügge who then published a more scientific article in *Die Naturwissenschaften* in June 1939 and 2 months later a popular article in the *Deutsche Allgemeine Zeitung* on the possibilities of exploiting atomic energy, thus drawing the attention of a wider public to the new possibility of energy production.[64]

Niels Bohr left Copenhagen at the beginning of January 1939 for a research stay of several months in Princeton, USA. A few days before his departure, Frisch returned from Sweden and informed Bohr about the latest findings.[65] On January 16, 1939, Bohr arrived in New York, i.e., even before the said issue of *Die Naturwissenschaften* arrived there. Bohr had promised Frisch that he would not make any announcements before the corresponding publications appeared, but as it is, Bohr—who of course was concerned with this sensational discovery during the entire sea-crossing—had already told John Archibald Wheeler (1911–2008), who had received him, the news on the quay under the seal of secrecy, and—as is usual with such confidential information—it spread like wildfire.[66] As surprising and

---

[62] In Vienna, it was Josef Schintlmeister (1908–1971) and Werner Czulius (1913–2008) who made corresponding considerations in Jan/Feb 1939 and presented them to the Vienna Academy on February 23, 1939. [Cf. Nagel, G.: Atomversuche in Deutschland. Zella-Mehlis: Heinrich Jung Verlagsgesellschaft 2003, p. 24 f].

[63] In 1949 Flügge remarked somewhat smugly: "When we returned from the Christmas vacations at the beginning of 1939, everything had already been decided." (Flügge, S.: Zur Entdeckung der Uranspaltung vor zehn Jahren. In: Zeitschrift für Naturforschung 4a(1949), pp. 82–84). According to the entry in his pocket diary, Hahn had discussed the discovery with von Weizsäcker, Flügge and others on Monday, January 9, 1939.

[64] Flügge, S.: Kann der Energieinhalt der Atomkerne technisch nutzbar gemacht werden? In: Die Naturwissenschaften 27(1939)23/24 (published on 9.6.1939), pp. 402–410; Flügge, S.: Die Ausnutzung der Atomenergie. In: Deutsche Allgemeine Zeitung of August 15, 1939. At the same time (on June 16, 1939) Georg Stetter (1895–1988) from Vienna applied for a patent at the German Reich Patent Office on how nuclear fission can be used to generate energy (cf. Nagel, G.: Atomversuche in Deutschland. Zella-Mehlis 2003, loc. cit., p. 29).

[65] Frisch, O. R.: What little I remember. Cambridge University Press 1979, p. 116; Röseberg, U.: Niels Bohr. Leben und Werk eines Atomphysikers. Berlin: Akademie-Verlag 1985, p. 215.

[66] The records are not entirely clear on whether Bohr had already informed Wheeler and possibly Fermi, who had also welcomed him on the quay, that same evening; certainly, however, he had not told his accompanying assistant Leon Rosenfeld (1904–1974), with whom he had discussed the matter during the passage of the ship, that silence was still to be maintained, and Rosenfeld

unexpected as the discovery had been, after it had become known everyone working in the field of atomic research was immediately aware not only of the significance of this discovery, but it could also be understood relatively quickly, both experimentally and theoretically.

Consequently, Bohr had no choice but to officially report on this at the Washington Conference on Theoretical Physics, which that year was already taking place at the end of January and was actually concerned with low-temperature physics.[67] However, this meant that the attention paid to the Meitner–Frisch publication a few weeks later remained low, and Lise Meitner's contribution to this work in particular was therefore initially underestimated in the scientific world.

In the following weeks Bohr continued to work with Wheeler on the theory of nuclear fission, and their fundamental article appeared in *Physical Review* on September 1, 1939—a date which, as is well known, is also of fundamental political and historical importance—namely the outbreak of the Second World War.

The global political situation in the spring of 1939 led to the consequence that physicists all over the world were not only enthusiastic about the new findings and fascinated by the possibility of almost unlimited energy production; they also recognized that this release of nuclear energy held the possibility of immense destruction and was therefore a candidate for a new type of bomb. In this new type, however, scientists and military officers alike saw merely a "common" bomb with "only" much greater destructive power—the fact that this was to be an absolutely new type of weapon, which would also require new military–political thinking, became clear to most of them only after the first terrible use over Japan.[68]

In my opinion, the relatively short period of time between the discovery of nuclear fission in mid-December 1938 and the publication of their theory at the beginning of September 1939 is an essential aspect for discussing the ambivalence of this scientific result. All the scientists involved were obviously aware of this ambivalence, but none of them drew the conclusion that they should not continue to participate in this research. Some scientists were considering stopping publication, whereas others opposed this as a threat to the freedom of basic research. Thus, Szilard already asked Frédéric Joliot-Curie in early February 1939 whether he would join a voluntary embargo on further publications, for he was aware of Joliot-Curie's research on neutron multiplication.[69] Joliot-Curie replied that although he

divulged the news to Wheeler as well as to the Princeton club of scientists that same evening: See Röseberg, U.: Niels Bohr; loc. cit.; also Wheeler, J. A.: Geons, Black Holes & Quantum Foam. A Life in Physics. New York/London 1998, Chapter 1.

[67] Interestingly, although Bohr's remarks are mentioned on the corresponding commemorative plaque at George Washington University, the names of Hahn, Straßmann and Meitner are not mentioned. Cf. https://physics.columbian.gwu.edu/1939-fifth-washington-conference-theoretical-physics-low-temperature-physics-and-superconductivity (accessed: 10 April 2018).

[68] Cf. inter alia Kant, H.: J. Robert Oppenheimer. Leipzig: Teubner 1985.

[69] Goldsmith, M.: Nuclear Fission and War. In: New Scientist (1976, June 17), pp. 646–647.

agreed in principle with Szilard's intention, it could not be assumed that all laboratories would adhere to it, and therefore he too would continue to publish.[70]

From mid-1940 onwards, however, nothing more appeared in American or other journals on this subject, particularly given the understandable rationale that no further information should be made available to the Germans.[71] However, especially in Germany and the Soviet Union—i.e., countries at a comparable level of research at that time—it was naturally concluded from this hiatus in publication that the USA was working on an atomic bomb.[72] The question arises of whether the pause in publication really had a "positive" effect.

Especially the scientists who emigrated from Germany and Italy before fascism soon realized the danger posed by an atomic bomb in the hands of the German fascists. This was true for those who emigrated to the USA as well as to England and France, and they endeavored—sometimes in cooperation with scientists from those countries—to make the respective governments aware of this danger.

Leo Szilard, who had worked closely with Einstein and others in Berlin in the 1920s and who, as mentioned above, had already patented the principle of the nuclear chain reaction in 1934, made a first attempt—without success—in March 1939, together with Fermi and others, to gain the interest of American government agencies. In order to make a second attempt more successful, Szilard assured himself of Einstein's support in the summer of 1939, which is how the famous letter of August 2, 1939, written by Szilard and signed by Einstein, to U.S. President Franklin D. Roosevelt (1882–1945) came about. From this—to put it very simply—the US American atomic bomb program finally resulted, which, however, was not specifically coordinated by the state until the end of 1940 and did not even begin in its full complexity until the summer of 1942, when the military took over the organizational management and overall responsibility, then under the codename of the *Manhattan Project*. How would Szilard, Einstein, and others view this effort from today's perspective; what should they possibly have done differently?

The fears of those scientists were well founded. By the end of April 1939, a number of leading German nuclear researchers had already gathered in the German *Reichserziehungsministerium* [Ministry of Science, Education and National Culture] to coordinate the relevant research. After the beginning of the war, Germany's most important nuclear researchers were called to the *Heereswaffenamt* [Army Ordnance Office], including Hahn, Heisenberg, and von Weizsäcker. Unofficially, this group was referred to as the *Uranverein* [Uranium Association],

---

[70] Ibid., p. 647.

[71] In general, the considerations regarding the publication suspension only referred to the possible military use, i.e., to the actual "abuse" of such technology. But at that time, this was ultimately only seen in the case of the political–military opponent (to whom no information was therefore to be given), whereas their own military application was approved, as the following comments show.

[72] In addition, there was corresponding secret service information, for example in the Soviet Union (cf: Atomnyj proekt SSSR. Documenty i materialy. Tom 1.1 (1938–1945), Moskva 1998; various materials, including pp. 121–122 the letter from three Academy members of June, 12 1940 to the Deputy Chairman of the Council of Ministers of the USSR).

and at the opening of this project it was made clear that the question of the technical use of nuclear energy was not only of general military interest, but was also directly related to possible new weapons.[73] The first thing Heisenberg was asked to do was to work out the theoretical principles of a nuclear reactor—in Germany at that time they called it an *Uranmaschine* [Uranium engine]. In this context, it is important to emphasize that this military uranium project was not at first explicitly a bomb program, although the background was clear.

In a 1984 interview about the German uranium project, Weizsäcker also asks the question, what would we have done if we had realized that we could make the bomb? "That we said: Dear Führer, we know how to make the bomb, but we will not tell you? I'd like to see how we would have done that."[74]

In the spring of 1942, when decisions were due on whether to continue the German project, it did not become a direct atomic bomb program. This had nothing to do either with the sometimes-alleged incompetence of the scientists involved or with possible ethical and moral concerns. The reasons behind this were mostly due to other political, military, and economic contexts. Consequently, from summer 1942 onwards, the German uranium program ran largely as a more "civilian" program for the construction of a uranium reactor. Nevertheless, interested parties from the Wehrmacht and SS had not abandoned the idea of an atomic bomb.[75]

Ten of the leading German nuclear scientists—among them Hahn, Heisenberg and Weizsäcker—learned about the first use of atomic bombs by the USA over Japanese cities after the European end of the Second World War at the British Farm Hall, where they spent 6 months in Western Allied captivity, as the American secret service in particular wanted to get an idea of what progress the Germans had actually made in the development of atomic bombs and whether they could still be used productively for the Manhattan Project. This is not the place to discuss this aspect in more detail. As a result of the protagonists' conversations at Farm Hall, their individually diverse entanglements in the German uranium project became the collective experience of an illegitimate appropriation of science by political power. From the retrospective construction of resistance to this use would soon arise—in connection with the rearmament of West Germany after the war—the further potential of real resistance amongst scientists to such indoctrinations.

---

[73] Bagge, E., Diebner, K. & Jay, K.: Von der Uranspaltung bis Calder Hall. Hamburg: Rowohlt 1957, p. 23.

[74] Die Atomwaffe. Interview 1984 with H. Jaenecke from *Stern*. Reprinted in: von Weizsäcker, C. F.: Bewußtseinswandel. München/Wien: Carl Hanser 1988, pp. 362–383 (here p. 367). (translation by the author)

[75] Numerous studies on the German nuclear project from recent years make this clear. Cf. among others: Karlsch, R.: Hitlers Bombe. Die geheime Geschichte der deutschen Kernwaffenversuche. München: Deutsche Verlags-Anstalt 2005; Nagel, G.: Atomversuche in Deutschland. Geheime Uranarbeiten in Gottow, Oranienburg und Stadtilm. Zella-Mehlis/Meiningen: Heinrich-Jung-Verlagsgesellschaft mbH 2003; Nagel, G.: Wissenschaft für den Krieg. Die geheimen Arbeiten der Abteilung Forschung des Heereswaffenamtes. (= Pallas Athene, Vol. 43) Stuttgart: Franz Steiner 2012.

Hahn always emphasized that his research during the war had nothing to do with an atomic bomb, as only the products of uranium fission were studied at his institute. In principle, that's true, but a little naïve. For Hahn, for example, was a member of the Uranverein and knew—at least in principle—about the work, even on the assumption that he had relatively little understanding of the physical phenomena that were dealt with there (which is certainly not completely wrong). In autumn 1945 Hahn was awarded the 1944 Nobel Prize in Chemistry "for his discovery of the fission of heavy nuclei"; the award was presented at the Nobel celebration in December 1946, since Hahn was still interned at the British Farm Hall in 1945. Whether Meitner as well as Straßmann and Frisch should also have been included is another topic I have discussed elsewhere.[76]

Hahn, Meitner and Straßmann were then jointly awarded the Enrico Fermi Prize of the US Atomic Energy Agency in 1966.

***

Like most scientists of his generation, Hahn believed that the peaceful and military applications of scientific knowledge could be separated relatively clearly. He was convinced of the blessings of science for mankind, and so there was no question in his mind that the peaceful use of atomic energy was one of the most promising tasks for the future of mankind. At the end of his Nobel Lecture in December 1946, he therefore expressed himself very firmly in this sense:

> […] The energy of nuclear physical reactions has been given into men's hands. Shall it be used for the assistance of free scientific thought, for social improvement and the betterment of the living conditions of mankind? Or will it be misused to destroy what mankind has built up in thousands of years? The answer must be given without hesitation, and undoubtedly the scientists of the world will strive towards the first alternative.[77]

Due to the Allied Control Council Law No. 25, however, nuclear physics research was largely prohibited in post-war Germany.[78] However, the scientists around Heisenberg early on began to look for ways and means to mitigate or completely

---

[76] Cf. among others: Kant, H.: "… der Menschheit den größten Nutzen geleistet …"!? 100 Jahre Nobelpreis, eine kritische Würdigung aus historischer Perspektive. In: Physikalische Blätter 57(2001)11, pp. 75–79 (esp. pp. 77 f); Kant, H.: Die radioaktive Forschung am Kaiser-Wilhelm-Institut für Chemie von den Anfängen bis zum deutschen Uranprojekt. 2012, loc. cit. (here pp. 96–99).

[77] Hahn, O.: From the Natural Transmutations of Uranium to its Artificial Fission. (Nobel Lecture, 1946). In: Nobel Lectures, Chemistry 1942–1962, Elsevier Publishing Company, Amsterdam, 1964, pp. 51–66 (here p. 64).

[78] Allied Control Council Act No. 25 "Regulation and Supervision of Scientific Research" of 29.4.1946 [http://www.verfassungen.de/de45-49/kr-gesetz25.htm] (accessed: 4 October 2017), specified by Act No. 22 of the Allied High Commission "Supervision of Substances, Facilities and Equipment in the Field of Atomic Energy" of 2.3.1950.

circumvent this ban, and Hahn, by then president of the Max Planck Society, also supported these efforts.[79]

As far as the peaceful use of nuclear energy was concerned, in the 1950s, nuclear reactors based on the fission reaction had been developed to the extent that they could supply electrical energy. The first grid-connected power plant went online in 1954, and was located in Obninsk near Moscow. The fact that the Soviet Union was a pioneer in this field had more than just a political background. The Soviet Union was more dependent than the USA, for example, on developing new energy sources. On the other hand, it has to be said that this nuclear power plant was not a purely civilian foundation, but offered the possibility that some leading Soviet nuclear scientists, who were against military development for various reasons, could cleverly be involved in the relevant research. It had emerged from "Labor V" of the Soviet atomic bomb project, founded in 1945.

Britain and the USA followed a few years later, and here too the first nuclear reactors had emerged from military research. The development of the nuclear energy industry will not be further discussed here.

The nuclear weapons available in the 1950s were atomic bombs and hydrogen bombs, both fission bombs as well as fusion bombs. The international development of nuclear weapons was also regarded with concern by many people in post-war West Germany—especially with regard to the development of atomic and hydrogen bombs—but they were initially more concerned with their own reconstruction and also saw the nuclear weapons issue as a problem for the major powers. The coming international wave of protest at the beginning of the 1950s therefore hardly touched the Federal Republic of Germany. Activities such as those of the World Peace Council, founded in 1949 under Frédéric Joliot-Curie—for example the appeal to ban nuclear weapons, known as the Stockholm Appeal of 1950, to which millions of people around the world signed up—remained largely unnoticed in the FRG, due not least to fear of communist influence.

This changed around the turn of the year 1954/55, and at least three aspects are of particular importance at that time. Firstly, the Paris Treaties were signed in October 1954, which, among other things,[80] sealed the entry of the Federal Republic of Germany into NATO and thus the reconstruction of a German army. On the other hand, at the international level, the escalation of the Cold War and the consequences of the Korean War had led to a growing anti-nuclear war movement, which now reached Germany. And finally, the plans of the German nuclear scientists to be able to conduct nuclear research again within the "normal framework" came closer and closer to their realization, and the German scientists realized that it would be beneficial for their cause to make it clear that they were *only* interested in "peaceful nuclear research." That such peaceful nuclear research is beneficial to human society was a consensus not only among the Federal German scientists.

---

[79] The Control Council Law No. 25 was officially repealed for the FRG in May 1955 as a consequence of the Paris Treaties; in the GDR, it was subsequently repealed in September 1955.

[80] The Paris Treaties entered into force in May 1955.

In the spring of 1955, Hahn—inspired by correspondence with Max Born (1882–1970) and Bertrand Russell (1872–1970)—first held a radio lecture and then published a brochure entitled "Cobalt 60—Gefahr oder Segen für die Menschheit" [Cobalt 60—Danger or Blessing for Mankind]; Hahn had probably spoken more clearly than others before him about the dangers of misusing nuclear energy.[81]

In an exchange of ideas with Born, Heisenberg and Weizsäcker, he decided to take the initiative of using the Lindau Conference of Nobel Laureates, which has been held annually since 1951, to launch an appeal *against* military and *for* peaceful use of nuclear energy. Hahn succeeded in persuading all 16 Nobel Prize winners present in Lindau to sign.[82]

In connection with the question of equipping the West German Bundeswehr with nuclear weapons as a consequence of joining NATO, the so-called *Göttinger Erklärung* [Göttingen Declaration] was issued in 1957. The initiators here were, above all, Hahn and Weizsäcker. Important aspects of the Göttinger Erklärung were that the signatories were against nuclear armament of the Bundeswehr, that they made clear the dangers of nuclear weapons and that they declared that they would not participate in the production or testing of nuclear weapons; at the same time, they advocated research into the peaceful use of nuclear energy. This declaration was addressed exclusively to the German Federal Government, so it did not claim an international dimension like the *Mainauer Erklärung* [Mainau Declaration][83] and, for example, it explicitly did not oppose nuclear weapons tests.[84] An important result was the future participation of German scientists in the international Pugwash movement; for the anti-nuclear movement of the 1950/60s "on the street," however, the influence of the Göttingen Declaration was ultimately only slight, because the scientists involved were not prepared to "take themselves to the streets."

---

[81] Cf. Kant, H.: Otto Hahn und die Erklärungen von Mainau (1955) und Göttingen (1957). In: Vom atomaren Patt zu einer von Atomwaffen freien Welt. Zum Gedenken an Klaus Fuchs. (= Abhandlungen der Leibniz-Sozietät der Wissenschaften Vol. 32) Edited by Günter Flach & Klaus Fuchs-Kittowski. Berlin: trafo Wissenschaftsverlag 2012, pp. 183–197; See also: Kant, H.: Werner Heisenberg and the German Uranium Project. Otto Hahn and the declarations of Mainau and Göttingen. (= Preprint 203(2002) Max Planck Institute for the History of Science), esp. pp. 21–40. [https://www.mpiwg-berlin.mpg.de/sites/default/files/Preprints/P203.pdf]

[82] Cf. Hahn, O.: Mein Leben. op. cit. p. 230. In his notebook Hahn notes under 11.7.55 "In the afternoon even longer meetings with the 16 Nobel Prize winners present. Finally, Lipmann gives in." (in Hahn, D.: Otto Hahn – Begründer des Atomzeitalters. Eine Biographie in Bildern und Dokumenten. München: Paul List 1979, p. 249) (translation by the author); Cf. also Born's remark in: Max Born: My Life and My Views. New York 1968, p. 85. Of the first signatories, Compton and Yukawa were not present in Lindau (see figure of the original declaration in Hahn, D., loc. cit. p. 250; also in Gerlach, W. & D. Hahn: Otto Hahn – Ein Forscherleben unserer Zeit. (= Große Naturwissenschaftler Vol. 45). Stuttgart: Wissenschaftliche Verlagsgesellschaft 1984, p. 154 f).

[83] As *Mainauer Erklärung* [Mainau Declaration] is denoted the appeal appointed at the Lindau Conference of Nobel Laureates, which was made public during a meeting on the isle of Mainau.

[84] Hahn emphasized the latter explicitly in a letter to Karl Bechert (1901–1981) of December 12, 1957 [MPG Archive, NL Hahn, Abt. III, Rep. 14A, No. 00 208, p. 2]. Bechert was a physicist and SPD politician and dedicated himself early on to the fight against both military and civil uses of nuclear energy; later he was also a co-founder of the anti-nuclear power movement.

It is important to note that with the Mainau Declaration, and even more so with the Göttingen Declaration, leading German (natural) scientists moved from their immediate scientific sphere of activity to a broad political public for the first time. But did they really do so out of political conviction or rather for very different self-ish reasons? It is certainly reasonable to doubt whether this image-boosting appearance actually warrants the responsible political intent that has increasingly been attributed to it in recent years.

Hahn repeatedly emphasized that in this matter he was not acting as president of the Max Planck Society but as an independent scientist, but on the other hand he used the opportunity to represent the intensions of the *Göttinger Achtzehn* [Göttingen Eighteen] at the 1957 annual meeting of the Society in Lübeck at the end of his presidential address.[85]

There were different views among the signatories on the further course of action.[86]

For Hahn, the next suitable opportunity to bring his concern to the public was at his much-acclaimed lecture "Atomenergie für den Frieden oder für den Krieg" [Atomic Energy for Peace or War], which he gave on November 14, 1957 in the Vienna Konzerthaus, invited by the Austrian Cultural Association.[87] In this he remained true to his conviction that he was unequivocally opposed to nuclear war, and quoted the Mainau and Göttingen Declarations at length, but nevertheless vehemently advocated for peaceful uses of nuclear technology.[88]

Even Weizsäcker, who played a significant role in preparing the Mainau and Göttingen Declarations, was still guided by this conviction at that time. This is true even for the vast majority of scientists involved in the development of atomic bombs.

"[...] the physicists have known sin; and this is a knowledge which they cannot lose," commented J. Robert Oppenheimer (1904–1967), the so-called "father of the American atomic bomb" following the deployment of the two atomic bombs against Japan and after the end of the Manhattan Project in 1947. Oppenheimer did not define this "sin," but was clearly referring to military applications, and it must be noted that most of the scientists involved must be given a high degree of responsibility. They had decided to take this step precisely because they feared that this weapon would be unscrupulously abused by others who might possess it. Given their political and social backgrounds, they could hardly have guessed in those days that the

---

[85] Hahn, O.: Ansprache des Präsidenten auf der Hauptversammlung der MPG 1957 in Lübeck. In: Mitteilungen aus der Max-Planck-Gesellschaft, issue 4/1957 (August), pp. 194–201 (here p. 199 f).

[86] However, in accordance with the compromise agreed in the Federal Chancellery on April 17, 1957, the scientists were generally of the opinion that they should not go public before the federal elections in September 1957, in order not to expose themselves to the accusation of taking a party-political stand. See Kraus, E.: Von der Uranspaltung zur Göttinger Erklärung. Otto Hahn, Werner Heisenberg, Carl Friedrich von Weizsäcker und die Verantwortung des Wissenschaftlers. Würzburg: Königshausen & Neumann 2001; Rese, A.: Wirkung politischer Stellungnahmen von Wissenschaftlern am Beispiel der Göttinger Erklärung zur atomaren Bewaffnung. Frankfurt am Main: Lang 1999.

[87] Hahn accepted the invitation on 11.9.1957 [MPG Archive NL Hahn, Dept. III, Rep. 14A, No. 05 548, p. 15].

[88] Lecture manuscript in the MPG Archive NL Hahn, Dept. III, Rep. 14A, No. 06 379.

U.S. Government, which originally had to be persuaded by them to take up the development of atomic bombs, would later unscrupulously abuse this weapon itself.[89] But the peaceful use, which they advocated with varying degrees of vehemence in the 1950s and 1960s, was an undoubtedly positive option for them.[90]

**

It should not be forgotten that the euphoria among scientists regarding the peaceful use of nuclear energy in the 1950s and 1960s was supported by a political euphoria (not only in Germany), which was related on the one hand to the circumstances of the Cold War and on the other hand to economic considerations, which Franz Josef Strauß (1915–1988), Federal Minister for Atomic Questions from 1955–1956, put in a nutshell:

> But we must now take the first steps, quite modestly and simply, that are necessary for us to occupy an equal place in the circle of nuclear powers […] that use this power for peaceful purposes in the foreseeable future.[91]

---

[89] Cf. inter alia Kant, H.: J. Robert Oppenheimer. op. cit. p. 156; Bird, K. & Schweber, M.: American Prometheus – The Triumph and Tragedy of J. Robert Oppenheimer. New York: Alfred A. Knopf, 2005, esp. p. 388.

[90] Nevertheless, the United States' Acheson–Lilienthal Plan of 1946, a first draft for the international control of atomic weapons and the atomic energy industry, which had been drawn up with Oppenheimer's decisive involvement, had already made it clear: "The development of atomic energy for peaceful purposes and the development of atomic energy for bombs are in much of their course interchangeable and interdependent." [A Report on the International Control of Atomic Energy; Prepared for the Secretary of State's Committee on Atomic Energy. Washington, D.C. March 16, 1946, p. 10]. This basic problem could not be solved even by a clear commitment to peaceful use, as further developments showed, even though efforts were made to keep the military aspect in the background for the public.
  Hahn, for example, although he had again vehemently advocated the peaceful use of nuclear energy at the General Assembly of the Max Planck Society in Hannover in 1958, also expressed initial cautious doubts about this use. "For years I have occasionally wondered whether it wouldn't be better if the entire utilization of atomic energy had never become a fact […]," but then emphasized his positive view that it is beneficial to have atomic energy available in view of the shortage of coal and oil. However, he also points out the dangers of the fission products produced in the nuclear reactor as well as their use (e.g., plutonium) for the construction of atomic bombs, and then finally refers to the emerging possibility of nuclear fusion, emphasizing among other things that the fusion reactor cannot be used for the production of hydrogen bombs, but could solve the energy problem. [Hahn, Otto: Ansprache auf der Hauptversammlung der Max-Planck-Gesellschaft 1958 in Hannover. In: Mitteilungen aus der Max-Planck-Gesellschaft zur Förderung der Wissenschaften (1958) 4, pp. 216–224 (here pp. 221 ff)]. (translation by the author)

[91] Quoted after Fischer, P.: Atomenergie und staatliches Interesse: Die Anfänge der Atompolitik in der Bundesrepublik Deutschland 1949–1955. (= Internationale Politik und Sicherheit, Vol. 30/3) Baden-Baden: Nomos 1994, p. 261. (translated by the author). Fischer further shows that Strauß was by no means only interested in peaceful use, for he was convinced that "what the Bundeswehr was in one area, nuclear energy was in another" (ibid., p. 262). And finally, this military background also had an influence on the choice of fuel (natural uranium or enriched uranium) and thus on the reactor type [cf. e.g., Radkau, J. (1983) loc. cit., p. 63: "The fact that the Federal Republic of Germany refrained from building nuclear weapons did not prevent the development of nuclear energy from being pre-structured by military technology in this case, too"].

One of the consequences of this objective was that the German Government urged the energy sector to enter into nuclear power generation and announced that it would provide the necessary capital, because it was clear that the capital expenditure requirements would be much higher than for conventional power plants. The concerns of the energy industry were therefore mainly based on financial circumstances and less on risks of technical feasibility or ecological considerations.[92] And nothing much has changed since then.

Simultaneously, scientists and politicians at that time repeatedly depicted an international race in which countries found themselves with regard to the use of nuclear energy and in which they should not be left behind. Erich Bagge (1912–1996) opposed this as early as 1960—probably rather unconsciously, because it seemed natural to him:

> The question of whether it would be appropriate at all to get involved in this particularly expensive race for the peaceful use of nuclear energy was not at any time a problem for discussion.[93]

The concept of the peaceful use of nuclear energy was not limited to the generation of energy for public power grids by nuclear power plants. There were also numerous proposals to use nuclear energy directly for propulsion technology, for example for ships.[94] However, military use was also a pioneer in ship propulsion: in 1955, the first US nuclear submarine went into operation.[95] In Germany, the Society for the Utilization of Nuclear Energy in Shipbuilding and Shipping (GKSS, *Gesellschaft*

---

[92] Cf. i.a. Radkau, J.: Aufstieg und Krise der deutschen Atomwirtschaft 1945–1975. Verdrängte Alternativen in der Kerntechnik und der Ursprung der nuklearen Kontroverse. Reinbek bei Hamburg: Rowohlt 1983, p. 196 ff; Müller, W. D.: Die Geschichte der Kernenergie in der Bundesrepublik Deutschland. Vol. I: Anfänge und Weichenstellungen. Stuttgart: Schäffer-Poeschel 1990 (esp. Chapter B.6.: Wer braucht Kernenergie?). On the corresponding development in the GDR, see for example Liewers, P., Abele, J. & Barkleit, G. (Eds.): Zur Geschichte der Kernenergie in der DDR. Frankfurt am Main: Peter Lang 2000.

[93] Bagge, E.: Die friedliche Nutzung der Kernenergie für technische Zwecke. In: Atom. Wirklichkeit – Segen – Gefahr. Werbebroschüre des Innenministeriums des Landes Schleswig-Holstein, Kiel 1960, pp. 55–65 (here p. 56). Bagge as well as Kurt Diebner (1905–1964) were among the members of the *Uranverein* who were detained at Farm Hall in 1945. However, in many nuclear–physical questions they differed from the circle around Heisenberg, which had not only to do with the fact that during the war they conducted their nuclear–physical research in the research center of the *Heereswaffenamt*. Bagge, a student of Heisenberg, worked at the University of Hamburg from 1948 to 1957 and was then head of the Institute for Pure and Applied Nuclear Physics at the University of Kiel. Together with Diebner he was also one of the founders of the GKSS mentioned below.

[94] For this and other examples of use, see Margulies, R.: Atome für den Frieden. Köln/Opladen: Westdeutscher Verlag 1965. Although this book was written with enthusiasm for nuclear energy in mind, it also mentions, for example in the section "The Search for Energy Sources" (pp. 13–18), a wide variety of alternative energy sources, including wind and solar energy, which could not be reliably used in industry in those days due to the technological limitations of the time.

[95] At present, the USA, Russia, France, Great Britain, the People's Republic of China, and India operate submarine fleets powered by nuclear energy. Nuclear-powered aircraft carriers have been built in the USA since 1960; there are plans to build such carriers in Russia and China.

für Kernenergieverwertung in Schiffbau und Schiffahrt), based in Geesthacht, published a tender for a nuclear-powered merchant vessel in 1960,[96] which was launched in 1964. In 1968 the ore carrier *Otto Hahn* was commissioned.[97] The nuclear merchant fleet was therefore limited to this one ship, which was decommissioned in 1979 and scrapped in 2009.[98] Nuclear cargo ships were not able to establish themselves for various reasons (above all, they turned out to be uneconomical).

Obviously, aspects which today play a major role in the assessment of the peaceful use of nuclear energy initially played only a minor role or none at all. The main focus was on the possibility of relatively unlimited availability of energy (which was also considered "clean" compared to the limited fossil fuels) and the beneficial use of radioactive isotopes, for example in medicine and technology. Although reactor safety was an important issue from the very beginning, the problem of, for example, the long-term safe storage of so-called nuclear waste (such as fuel elements, construction waste from nuclear reactors, etc.) or the technical, financial and time expenditure for the demolition of reactors initially played a subordinate role or was regarded as relatively unproblematic to solve, probably also due to the assumption that nuclear fission energy would soon be replaced by fusion energy. However, when fusion research began at the end of the 1950s it was assumed that a fusion reactor could be realized within 20–30 years at most, whereas today it is assumed that commercially viable nuclear fusion cannot be expected before 2050 (i.e., more than 100 years after the initial research began).[99] And even when it comes to the question of whether fusion energy is so much "cleaner" or more environmentally friendly than fission energy, estimates today are not as optimistic as they were at the end of the twentieth century.

In the mid-1980s, however, Weizsäcker wrote retrospectively in the introduction to a study by Meyer-Abich and Schefold, concluding that nuclear energy—in comparison to solar energy and technically feasible energy savings—did not sufficiently meet the criteria of social compatibility or economic viability:

> According to my scientific background, I was a spontaneous supporter of nuclear energy until the early seventies. [...] In the winter of 1974–75, as advisor to the Federal Minister for Research and Technology, I pointed out the inevitable coming public debate on nuclear energy. The form in which the public criticism of nuclear energy was then voiced was, of course, in my opinion, far too undifferentiated in terms of the facts.

---

[96] This was preceded by the Soviet nuclear icebreaker *Lenin* (launched in 1957, commissioned in 1959) and the US merchant and passenger ship *Savannah* (planned since 1955, commissioned in 1962).

[97] Hahn participated in the launch on June 13, 1964 in Kiel and saw this project as a confirmation of his ideas on the peaceful use of nuclear energy (see Hahn, D., loc. cit. p. 324; Hoffmann, K., loc. cit. p. 246).

[98] In 1979 the nuclear reactor was replaced by a diesel engine; in 1983 the ship was also converted into a container ship and continued to sail under changing names and flags until it was scrapped as a cargo ship. The removed reactor was stored at GKSS until 2010. Cf. e.g., Neumann, H.: Vom Forschungsreaktor zum "Atomschiff" Otto Hahn: Die Entwicklung von Kernenergieantrieben für die handelsmarine in Deutschland. (= Deutsche Maritime Studien 7) Bremen: H. M. Hauschild 2009.

[99] Radkau even calls the fusion reactor the "Fata Morgana der Atomeuphorie" [mirage of atomic euphoria]. [Radkau, J. & L. Hahn: Aufstieg und Fall der deutschen Atomwirtschaft. München: oekom 2013, p. 53].

[…] I now strongly support solar energy as the main energy source, supported by technically possible energy savings, and oppose the choice of nuclear energy as the main energy
source; […].[100]

Radkau commented on this change of opinion by a leading nuclear physicist of the
pioneering era with the following words:

[…] Weizsäcker had become skeptical of nuclear power, and that was a deep shock for them
[meaning the nuclear community – HK]—as if they had lost their demigod. A remarkable
thing: nuclear power is based on the laws of physics and a huge army of experts, and yet its
promotion was so dependent on the blessing of key personalities.[101]

From today's perspective, Weizsäcker's statement can be generalized to the effect
that there are no serious long-term alternatives to renewable energies.[102] In this
assessment, the danger of nuclear catastrophes plays a role, along with other aspects,
as does the still unresolved issue of nuclear waste disposal.[103] However, if in the
1950s and 1960s—depending on one's point of view—a certain nuclear hysteria or
euphoria was observed in science and politics, then today, on the other hand, in the
discussions about the relatively sudden nuclear phase-out after the Fukushima
catastrophe in March 2011, there is a danger of lapsing into anti-nuclear hysteria or
euphoria.[104] Although a nuclear phase-out had already been in the sights for some

[100] von Weizsäcker, C. F.: Introduction to Klaus Michael Meyer-Abich & Bertram Schefold: Die
Grenzen der Atomwirtschaft. München: C. H. Beck 1986, p. 15 f. (translation by the author)

[101] Interview by Frank Uekoetter with Joachim Radkau. In: Environmental History 13(2008)4,
p. 757–768 (here p. 760).

[102] Hubert Laitko summarizes Weizsäcker's reflections on these issues in this way: "When weapons
are developed, their destructive effect is consciously taken into account; it is the genuine purpose
of weapons to achieve precisely this effect. Ecological damage, on the other hand, is usually unintentional and even contrary to the objectives of the actions that cause such damage. Cognitively
and ethically, we are dealing with a completely different situation here than in weapons development. The actual complication is that here the damage does not unintentionally occur instead of the
intended benefit, but the intended benefit is achieved and in *addition* – as a secondary or consequential effect – there is also ecological damage." [Laitko, H.: Der Ambivalenzbegriff in Carl
Friedrich von Weizsäckers Starnberger Institutskonzept. Max Planck Institute for the History of
Science, Preprint 449, Berlin 2013, p. 16]. Basically, since the 1960s/1970s such considerations
have led to the establishment of the research field of *technology assessment*.

[103] An overview of nuclear accidents is provided by Mahaffey, J.: Atomic Accidents. A History of
Nuclear Meltdowns and Disasters. From the Ozark mountains to Fukushima. New York/London:
Pegasus Books 2015.

[104] In the meantime, there are numerous publications on the issues and consequences of the so-
called energy transition and the nuclear phase-out. In connection with what is presented here I only
refer to the following books: Banse, G. et al. (Ed.): Energiewende – Produktivkraftentwicklung
und Gesellschaftsvertrag. (= Abhandlungen der Leibniz-Sozietät, Vol. 31) Berlin: trafo
Wissenschaftsverlag 2013; Ostheimer, J. &. Vogt, M. (Eds.): Die Moral der Energiewende.
Risikowahrnehmung im Wandel am Beispiel der Atomenergie. (= Ethik im Diskurs, Vol. 10)
Stuttgart: W. Kohlhammer 2014; Morris, C. & Pehnt, M. (2016): Energy Transition. The German
Energiewende. Heinrich-Böll-Stiftung 2016. [https://book.energytransition.org/sites/default/files/
downloads-2016/book/German-Energy-Transition_en.pdf] (Accessed: 14 February 2018);
Kemfert, C.: Das fossile Imperium schlägt zurück. Warum wir die Energiewende jetzt verteidigen
müssen. Hamburg: Murmann 2017.

time in the Federal Republic of Germany,[105] it was not until 2000 that the red–green government under Gerhard Schröder made a corresponding decision. In 2009, key points of this proposed phase-out were effectively rescinded by the CDU/CSU/FDP government under Angela Merkel (nuclear energy should again be regarded more strongly as a "bridging technology").[106] However, following the Fukushima reactor disaster, the same Merkel government then abruptly made an unprecedented turn-around by deciding to phase out nuclear power, which now provides for the shut-down of all German nuclear power plants by 2022. At the same time, this decision manifested publicly in the so-called *energy transition*, which means the transition from fossil fuels and nuclear energy to a "sustainable" energy supply through so-called renewable energies.[107, 108] The problems of nuclear phase-out and energy tran-sition must be considered not least in connection with the problems associated with climate change.[109]

Internationally, however, the phasing out of nuclear energy is far from being a consensus, but is the subject of highly controversial discussions.[110] Just recently, various worldwide efforts to develop and build new generations of nuclear power

---

[105] The term "nuclear phase-out" emerged as a catchword of the anti-nuclear movement in the 1970s. On the origins of the anti-nuclear movement, see Radkau/Hahn, op. cit. p. 288 ff.

[106] This was due not least to the myth of the so-called "nuclear renaissance," which was propagated for the last years of the twentieth century and the first decade of the twenty-first century (the term was probably coined by William J. Nuttall); the first nuclear reactors of the so-called third genera-tion went into operation. For a concise overview of the time in 2011, see e.g., Joskow, P. L. & Parsons, J. E.: The Future of Nuclear Power After Fukushima. In: Economics of Energy & Environmental Policy 1(2012)2, pp. 99–113 (esp. pp. 101–105).

[107] The commonly used term "renewable energy" is actually misleading, because energy cannot renew itself (law of energy conservation!). It would be better to speak of "regenerative energy" (where "regenerative" refers to the energy source).

[108] In Germany, the debate on the energy transition began as early as in the 1970s/1980s within the environmental movement and with the entry of the Green Party into the Bundestag. (see Morris/Pehnt loc. cit., p. 55 ff). Since 2011, the word "energy transition" has increasingly been used as a synonym for nuclear phase-out, which, however, tends to conceal the connections. Although the nuclear phase-out is part of the energy transition, it is only one part of it [Kemfert loc. cit., chapter Postfakt 1].

[109] Cf. for example Edenhofer, O. & Jakob, M.: Klimapolitik. Ziele, Konflikte, Lösungen. (= C. H. Beck Wissen bw 2853) München: C. H. Beck 2017.

[110] Cf. inter alia the strategy paper PINC of the EU Commission of April 2016: [https://ec.europa.eu/transparency/regdoc/rep/1/2016/DE/1-2016-177-DE-F1-1.PDF],   [https://ec.europa.eu/ger-many/news/kommission-veröffentlicht-bericht-zur-kernenergie-der-eu_de].   (Accessed: 25 February 2018). In Germany, the Federal Government reacted with protest to these plans by the EU Commission [http://www.spiegel.de/wirtschaft/soziales/atomkraft-strategie-der-eu-bundesregier-ung-reagiert-empoert-a-1092664.html]. (Accessed: 25 February 2018).

plants have become known.[111, 112] At the same time, however, it should not be under-estimated that the presumably greater danger still comes from the military use of nuclear energy, not only with regard to its possible use in armed conflicts,[113] but also with regard to the disposal of both obsolete nuclear warheads and similar devices and also the corresponding infrastructure used to produce nuclear weapons.

This is not the place to discuss the pros and cons of a nuclear phase-out and energy transition, because this article was primarily concerned with scientific–historical considerations on the history of the discovery of nuclear energy in order to better understand this phenomenon and its significance for science and society. But on the other hand, scientific discovery cannot be separated from considerations of its use, as we have seen.[114]

This history of discovery also shows, however, that there is apparently always a certain euphoria at the beginning of such an application, and that risks and dangers

---

[111] In the international discussion on possible new nuclear power plants, the development of generation III and IV plants plays a decisive role (in this context the first generation refers to reactors built between 1950 and 1960, which are more likely to be understood as demonstration power plants, and the second generation refers to reactors built between 1970 and 1990, most of which are still in operation today). Key objectives in the development of the fourth generation are, among other things, increased safety, sustainable use of uranium and the recycling of waste (how far this is actually realistic is another question). Cf. among others [https://de.nucleopedia.org/wiki/Generation_IV], [https://www.heise.de/tr/artikel/Neue-Heimat-Kanada-3889812.html], [http://www.ageu-die-realisten.com/archives/2423] (accessed: 25 February 2018).

[112] The countries with the largest nuclear power capacity are currently (highest first): USA, France, China, Russia, South Korea, Canada (if we count "reactors in operation" we get: USA, France, China, Russia, Japan, South Korea). The USA, France, China, Russia and India (at least) are building or planning new reactors. See the Wikipedia article "Kernenergie nach Ländern" [https://de.wikipedia.org/wiki/Kernenergie_nach_Ländern] (accessed: 31 March 2021); see also: Nuclear Power in the World Today. [http://www.world-nuclear.org/information-library/current-and-future-generation/nuclear-power-in-the-world-today.aspx] (accessed: 31 March 2021). Thirteen EU Member States currently operate nuclear power plants (so approximately a quarter of the reactors operated worldwide are on EU territory). Austria, on the other hand, has never put its only nuclear power plant at Zwentendorf (project started in 1971, decommissioned in 1978) into operation; since 1999, the Bundesverfassungsgesetz für ein atomfreies Österreich [Federal Constitutional Law for a nuclear-free Austria] has been in force there. In Vienna there is only one research reactor still in operation. (Cf. inter alia Forstner, Christian: Zur Geschichte der österreichischen Kernenergieprogramme. In: Kernforschung in Österreich. Wandlungen eines interdisziplinären Forschungsfeldes 1900–1978. Edited by S. Fengler & C. Sachse. Böhlau: Wien etc. 2012, pp. 159–180). Turkey, on the other hand, is building its first nuclear power plant with Russian help in Akkuyu on the Mediterranean coast, which is scheduled to go online in 2023. Another nuclear power plant is being built in Sinop on the Black Sea coast with Japanese support (which seems to be stalled at the moment). Both regions are prone to earthquakes! [Wikipedia article "Nuclear energy in Turkey" (accessed 31 March 2021)].

[113] Cf. e.g., von Weizsäcker, E. U. & Wijkman, A.: Wir sind dran. Club of Rome: Der große Bericht: Was wir ändern müssen, wenn wir bleiben wollen. Eine neue Aufklärung für eine volle Welt. Gütersloher Verlagshaus 2017 (Chapter 1.6.2 Atomwaffen – die verdrängte Bedrohung).

[114] In this context, reference should also be made to a discussion event organized by the Leibniz-Sozietät on April, 12 2018: Die Energiewende 2.0: Essentielle wissenschaftlich-technische, soziale und politische Herausforderungen. [https://leibnizsozietaet.de/internetzeitschrift-leibniz-online-nr-29-2017/] (Accessed 29 August 2021).

only come into view later.[115] This probably also applies to the energy transition in the broader sense—many problems of wind or solar energy plants, for example, are currently being sidelined rather than being subjected to serious consideration for the future, and therefore there is a real danger that "mistakes" similar to those made in the introduction of nuclear energy will be made again.[116]

However, the question of how far research can or should actually go remains unanswered in this context. Walther Gerlach (1889–1979) already drew attention to the following connection in the mid-1950s:

> [...] It is, after all, the method of physical research that "new territory" is not discovered in nature, but is created in the artificially prepared world of the laboratory: by creating conditions that can be clearly overlooked and in which disturbing secondary conditions are eliminated as far as possible.[117]

This means, however, that the laboratory experiment always takes place under restricted, "artificial" conditions, i.e., it is actually an intervention in nature. Incalculable "unintended" consequences cannot be excluded.[118]

One aspect of this ambivalence is, among other things, the fact that in researching the transuranium elements, experiments were apparently carried out that do not occur in nature in this way, i.e., they represent an intervention in nature under laboratory conditions. But does the ambivalence mentioned here really only apply to Hahn's and Straßmann's experiments on transuranium elements, or does it already apply to the early 1930s with regard to the discovery of artificial radioactivity? Should the research have been stopped there already, since it was finally an intervention in nature? But is that really the case? In his 1939 contribution to *Die Naturwissenschaften,* Flügge had already raised the question: "If such a conversion of uranium is possible, [...] why did nature not anticipate this experiment and carry

---

[115] Other corresponding examples from the history of science and technology can be cited, for example the initially inflationary use of X-rays, the agricultural use of the insecticide DDT or the use of CFCs as propellants and refrigerants.

[116] A similar problem seems already evident in the so-called digitalization, which the German Federal Government formed in March 2018 has placed at the forefront of its socio-political objectives. When the newly appointed Minister of State for Digitalization, Ms. Dorothee Bär (CSU), picked up on an FDP slogan from the 2017 federal elections: "Digital first. Misgivings second" and shortly after taking office declared: "We also want to become digital world champion! [...], but I am tired of the misgivings" (cf: Interview with Dorothee Bär on March, 2018 [https://www.bild. de/politik/inland/dorothee-baer/flugtaxis-werden-in-wenigen-jahren-fliegen-55257916.bild.html], translated (accessed 8 April 2018)), then this is strongly reminiscent of the political approach to the introduction of nuclear energy in the 1950s and 1960s (cf. e.g., the statement by F. J. Strauß quoted above), and is at least a very questionable political understanding of how to deal with new scientific and technological developments, and makes a mockery of all historical experience, which is so readily emphasized especially with regard to nuclear energy. On the socio-political dimensions of digitalization, see e.g., Santarius, T. & Lange, S.: Smarte grüne Welt. Digitalisierung zwischen Überwachung, Konsum und Nachhaltigkeit. München: oekom 2018.

[117] Gerlach, W.: Die Kosten der modernen naturwissenschaftlichen Forschung. Mitteilungen aus der Max-Planck-Gesellschaft (1956) pp. 23–32 (here p. 29). Translation by the author.

[118] Parthey also makes explicit reference to this in his contribution to this book (see Ch. 7 of this volume).

it out in the rock?[119] At that time Flügge was unable to answer the question, but expressed suspicions about the possibility. In the meantime, however, a uranium ore deposit has been discovered in the Oklo region of Gabon in Africa, for example, where a reactor-like fission reaction took place about two billion years ago—only today there is no longer any fissile uranium remaining in the necessary concentration, so the process has stopped.[120]

How far may research actually go?[121] To what extent can researchers—especially basic researchers—be aware of consequences if they cross such boundaries? And where do these boundaries lie?[122] Can this be grasped by the term *ambivalence of science*? Not least, questions of scientific ethics also play an important role in physical and chemical research. The biophysicist and geophysiologist James Lovelock (*1919) regards the Earth, on the basis of his Gaia theory, as a kind of living organism![123]

The fact that Hahn and Straßmann finally discovered something other than what they had originally been looking for is ultimately also connected with this aspect of research, which cannot be discussed further here, however. On the other hand, it is not uncommon for research to reveal something that differs from expectations or even diverges entirely from the question originally investigated.[124]

---

[119] Flügge: Kann der Energiehaushalt […], op. cit., p 409.

[120] Meshik, A. P.: The Workings of an Ancient Nuclear Reactor. In: Scientific American (2005, November), pp. 83–91; Schaaf, M.: Kernspaltung im Herzen der Finsternis. Afrika und die Ursprünge des Nuklearzeitalters. In: "Radiochemie, Fleiß und Intuition – Neue Forschungen zu Otto Hahn." Ed. by Vera Keiser. Berlin, Diepholz: GNT Verlag 2018, pp. 433–476.

[121] In this context, Parthey refers to the German Embryo Protection Law in his contribution to this book (see Ch. 7 of this volume). Should something similar apply to research on non-living matter? At the very least, the validity of such a question cannot be denied, but it would require a fundamentally new understanding of physical and chemical experiments.

[122] Oppenheimer wrote in June 1946 in a kind of commentary on the Acheson-Lilienthal Plan: "The point is, "[…] that if you don't try to develop atomic energy, you can't control it—you can't say first we will control it, and then we will develop it, because the developmental functions are an essential part of the mechanism for control." [Oppenheimer, J. R.: The International Control of Atomic Energy. Bulletin of the Atomic Scientists 1(1946)12, pp. 1–5 (here p. 4)]. While this is quite understandable, it is ultimately a plea for not imposing any limits on (basic) research. Can this still be maintained nowadays?

[123] Lovelock, J.: The Revenge of Gaia. Why the Earth is Fighting Back and How We Can Still Save Humanity. Allen Lane, London 2006. (In German: Lovelock, J.: Gaias Rache. Warum die Erde sich wehrt. (= Ullstein Paperback 37210) Berlin: Ullstein 2008.)

[124] One speaks of serendipity; this term goes back to the sociologist of science Robert K. Merton (1910–2003). Often, people also talk about accidental discoveries [Zufallsentdeckungen], which on closer inspection are not so much "accidental" but instead require a precise ability to observe at the right time (often, it turns out afterwards, that others had already observed the phenomenon before the accredited discoverer, but did not interpret it correctly). As further examples we only refer to the discovery of X-rays or penicillin.

Chapter 7
Institutionalization, Interdisciplinarity, and Ambivalence in Research Situations

Heinrich Parthey

Abstract This article refers to the heart of science—research and the necessary freedom of research, which only arises with its institutionalization. Science deals with research problems in disciplinary and interdisciplinary research situations. An interdisciplinary research situation exists only when both the formulation of the problem and the methods used to address it require the participation of several disciplines. New scientific institutions, if nothing else, ensure interdisciplinary research. Since the times of Galileo Galilei, science has developed new, powerful research possibilities with the experimental method. The experimental method produces active changes that can also be accompanied by unintended consequences. Science must find a way to deal with this ambivalence without endangering its own freedom.

The emergence of new scientific disciplines and new forms of cooperation between their representatives for further progress in knowledge are two mutually dependent tendencies in the development of the sciences, both in their systems of order and in the research and teaching profiles based on them. These two tendencies can be used, above all, to grasp the change in the relationship between the object area of research and the object area of social practice, which conditions scientific disciplines as a form of historically established and changeable boundaries of knowledge and knowledge production. Ultimately, the view of scientific disciplines expressed by Max Planck as early as the 1930s applies here: "Their separation according to different subjects is not, after all, rooted in the nature of things, but arises only from the

---

This chapter was compiled from three contributions by Heinrich Parthey (2011, 2019, 2020), subsequently translated and edited. The line of argumentation created corresponds to the short, last contribution by Heinrich Parthey in the Wissenschaftsjahrbuch 2019 (Parthey, 2020), the latter being the basis for our book presented here.

H. Parthey (Deceased) (✉)
Gesellschaft für Wissenschaftsforschung GeWiF, Berlin, Germany
e-mail: vorstand@wissenschaftsforschung.de

141

limitations of human capacity, which inevitably leads to a division of labor" (Planck, 1944, p. 243, translated).

Scientific disciplines are historically conditioned and thus changeable forms of knowledge acquisition and knowledge reproduction in which both the manner of scientific questioning and the preference for certain methodological approaches are acquired and practiced by individual scientists, and in which scientists and scholars experience or can achieve social recognition and are institutionally established.

Based on the assumption that scientists have to refer to certain areas of theoretical knowledge both in formulating problems and in methically working on problems, a distinction can be made between disciplinary and interdisciplinary research situations.

Disciplinary and Interdisciplinary Research Situations

Science develops through theoretical thinking and observational activity, be it mere or experimentally determined observational activity, in which researchers methodically solve epistemological problems by means of knowledge and research techniques. Every problem refers to knowledge of situations in mental or observational or practical–experimental activity, in which the available knowledge is not sufficient to achieve the goals and therefore must be expanded accordingly.

In a narrower sense, the awareness of such a knowledge deficit is only called a problem if the missing knowledge cannot be taken over by others, but has to be gained anew. A *research problem* exists, if for a system of statements and questions about or according to conditions of goal achievement no algorithm is known, by which the determined lack of knowledge can be eliminated in a finite number of steps. If an algorithm is known, then a task exists. The conceptual differentiation between problem and task has also been fruitful in work for the methodology of modeling.

With the scientific problem the questions are justified by the existing knowledge, but not answered. A problem dissolves to the extent that new knowledge as substantiated information answers the questions that represent a scientific problem. An important difference exists between the occurrence of a problem situation, which is seized and represented by the researcher in the problem, and the existence of a research situation. Thus, the creative scientist must have a feeling for the really crucial questions, but at the same time (s)he must also have the correct feeling for it: to what extent it will at all be possible, with the given conditions of the research technology, to master the problems with the instruments available or which can be developed. According to this, a research situation can be understood as such connections between problem areas and *methodological structures* that allow the scientist to methically work on the problem areas by means of actual availability of knowledge and research technology.

Following the understanding of the methodological structure of research situations, in addition to the two entities of problem field and methodological structure

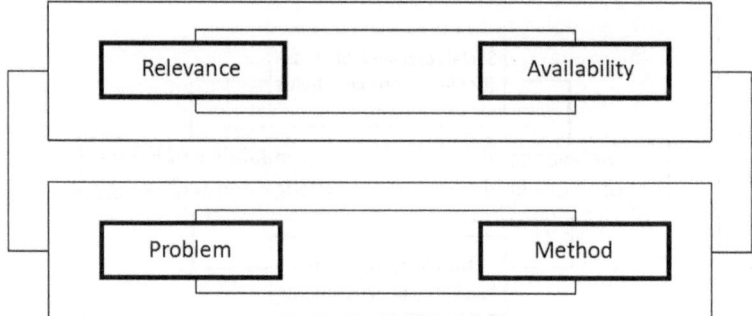

Fig. 7.1 Methodological structure of the research situation

and the relations between them, the actual availability of conceptual and physical means to deal with the problem on the one hand and the epistemological and social relevance of research problems on the other hand must also be taken into account (see Fig. 7.1). For if research situations with a novel relationship between problem and method as well as device (software and hardware) are to be brought about, only those research possibilities can be realized for which society provides the corresponding means and resources. Decisions on this, however, depend on the problem relevance shown. The *problem relevance*, i.e., the evaluation of the problems according to the contribution of their possible solution both for the progress of knowledge and for the solution of societal practical problems, ultimately regulates the actual availability of knowledge and equipment for problem processing.

At the end of the 1970s, Wolfgang Stegmüller, in discussion with Thomas Kuhn, attempted to define the concept of normal science more precisely with the help of the concept of *having a theory*. The term we use, *availability of knowledge and equipment for problem solving*, is much more comprehensive than the concept of having a theory at one's disposal, since it also includes practical feasibility in research. In a later version, for Stegmüller, "everything pushes towards a systematic pragmatics, in which non-logical terms are used, such as knowledge situation of persons and its change; subjective faith of persons at certain times; background knowledge available at a certain historical time and the like" (Stegmüller, 1983, p. 236, translated). In a further attempt in this direction, Stegmüller considers "additional pragmatic concepts that we have to build into the conceptual apparatus, because 'person', 'historical time', 'available knowledge', 'standards for the acceptability of hypotheses' are concepts of this kind" (Stegmüller, 1986, p. 109, translated).

If, in order to characterize research situations, the relationship between a problem field and a set of preconditions for problem solving is considered, then different research situations can be distinguished—at least according to the degree of the cognitive and social *relevance* of the respective problem and the degree of the actual *availability* of preconditions for solving the respective problem—but, above all, also according to their scientific and social integrity (see Fig. 7.2).

The relationship between the scientifically necessary disciplining of methodological problem-solving in research and the socially conditioned formulation of

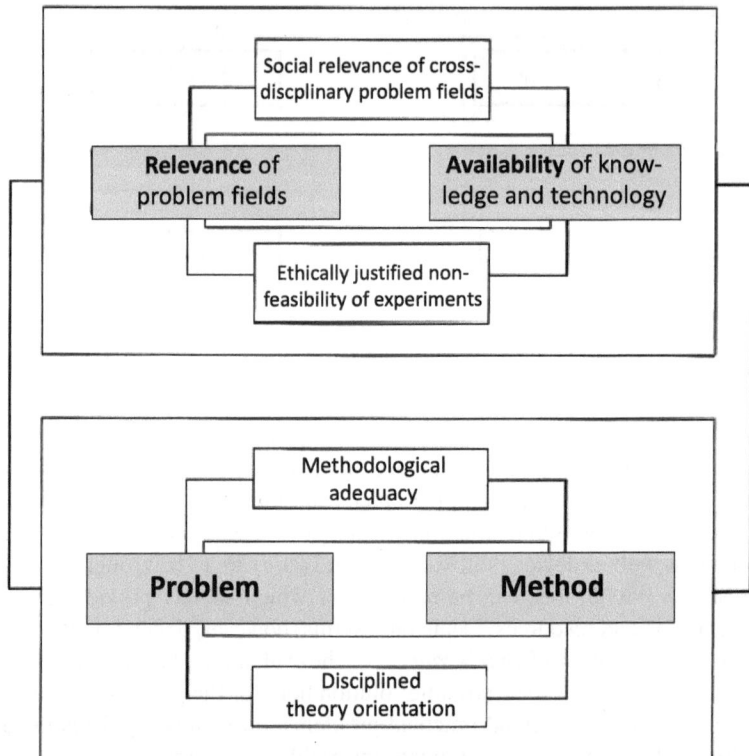

Fig. 7.2 Structure of scientific and social integrity of the research situation. (Bottom: scientific integrity, top: social integrity)

interdisciplinary problem fields for research leads to increased reflection on the distinction between disciplinary and interdisciplinary research situations: *A research situation is disciplinary if both the problems formulated in it and the methods used in it refer to one and the same field of theoretical knowledge, and a research situation is interdisciplinary if the problem and method of research are formulated or substantiated in different theories.*

Indicators of Interdisciplinary Work in Research Groups

As early as three decades ago, a comprehensive empirical study by UNESCO on the effectiveness of research groups asked, among other things: "In carrying out your research projects, do you borrow some methods, theories or other specific elements developed in other fields, not normally used in your research" (Andrews, 1979, p. 445). The first interpretations attempted to establish the comparability of the 1,200 groups studied by means of classification by discipline and interdisciplinary orientation in research. At the same time, it was assumed that the specific scope of cooperative relationships and thus co-authorship could be understood as a surrogate measure of the productivity of research groups working in interdisciplinary fields (cf. Steck, 1979).

The indicators of interdisciplinarity used by the current author in investigating 56 life science research groups in the years 1979–1981 were based on the assumption that the decisive factor for interdisciplinarity in research groups is whether *at least one group member* thinks in an interdisciplinary way, regardless of whether the group members are assigned to only one or several disciplines (cf. Parthey, 1982, 1983, 1990, 1997).

A first indicator of interdisciplinarity concerns the percentage of scientists in the research group who *formulate their problems in terms of (different) scientific disciplines in an interdisciplinary manner*. If all scientists in the group were to formulate problems in only one discipline, the percentage of scientists formulating problems across disciplines would be zero. Thus, groups that work on problem fields of the type mentioned are rightly classified as predominantly working in a *disciplinary* manner if, due to the derivation of sub-problems from a problem field, they are composed of representatives of different disciplines but work on these sub-problems via the means of their own discipline.

A second indicator of interdisciplinarity refers to the percentage of scientists in the group who need and use *methods to work on their problem that are not based in the same field of knowledge as the problem itself*. In this sense, our research enquired whether: "The methods used in the research group to work on your problem are: (A) grounded in the same area of knowledge in which your problem is formulated, [or] (B) are grounded in an area of knowledge that is different from the knowledge in which your problem is formulated." The percentage of scientists who answered with (B) in relation to the group size was recorded as the degree to which the *interdisciplinarity of problem and method* is expressed in research groups.

On the basis of these studies, the following forms of scientific activity can be distinguished (see Fig. 7.3):

1. Firstly, *monodisciplinary research* (i.e., in scientific activity, no cross-disciplinary problem has been formulated and no interdisciplinarity of problem and method has been developed).
2. Second, *multidisciplinary research* (i.e., in scientific activity, problems are formulated across disciplines, but no interdisciplinarity of problem and method has been developed).
3. Thirdly, *interdisciplinary treatment of disciplinary problems* (i.e., no interdisciplinary problem has been formulated in scientific activity, but interdisciplinarity of problem and method does occur).
4. And finally, fourthly, *interdisciplinary treatment of cross-disciplinary problem fields*.

Using the frequency of this combination of problem formulation in a cross-disciplinary context on the one hand, with the interdisciplinarity of problem and method on the other hand, we found the frequencies shown in Table 7.1. As we see,

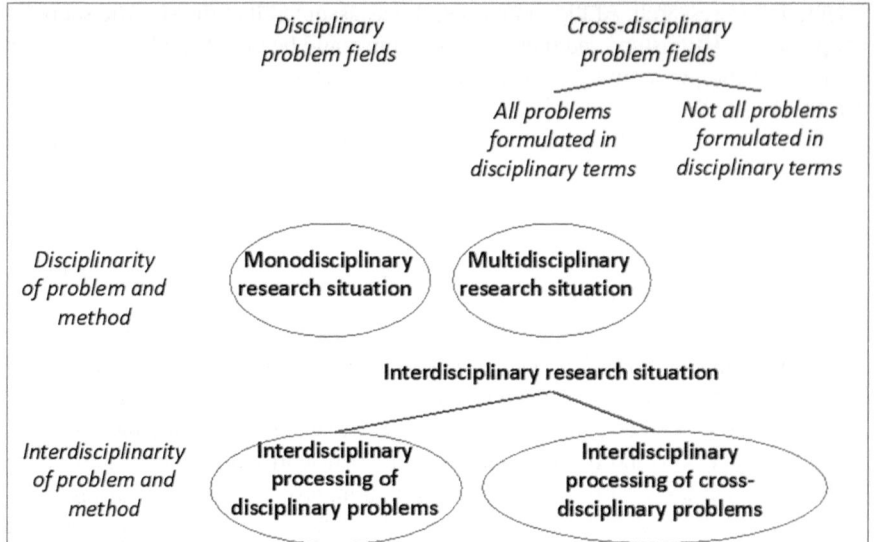

Fig. 7.3 Forms of scientific activity

interdisciplinarity of problem and method seems to be a necessary condition for full co-authorship in cross-disciplinary research groups.

Table 7.2 shows the frequency of the characteristic coupling between the indicator "Percentage of scientists working interdisciplinary with problem and method in research groups" and the indicator "Composition of research groups according to diploma disciplines." The first data column of Table 7.2 indicates that we find *personal interdisciplinarity, even if the research group is monodisciplinary* (i.e., the group members represent only one discipline). Interdisciplinarity and multidisciplinary composition of research groups do not coincide. In other words, *personal interdisciplinarity does not require multidisciplinary composition in the research group.* However, it can be assumed that the interdisciplinary work of individual scientists (understood as personal interdisciplinarity) is promoted by the composition of the research unit from representatives of different disciplines.

In addition to the indicators for interdisciplinarity mentioned above, further indicators were taken into account in analyses of interdisciplinary work, such as the indicator "co-authorship in the group," which is used as a surrogate measure of the productivity of interdisciplinary research groups and is based on bibliometric profiles of groups, as well as indicators for the "multidisciplinary composition of the group" (according to education) shown in Table 7.2 and the often used indicator for the "distribution of expertise by discipline." Our search for correlations between these indicators is based on an analysis of the rankings of the respective group values using correlation coefficients (see Table 7.3).

The many positive rank correlations—as shown in Table 7.3—can be interpreted to mean that a multidisciplinary training and competence structure of the group offers favorable conditions for interdisciplinary work by individual scientists.

Table 7.1 Interdisciplinary problem formulation and interdisciplinarity of problem and method in 56 research groups from four non-university institutes of the life sciences in the early 1980s

Cross-disciplinary problem fields	Interdisciplinarity of problem and method	# research groups	Chance of an encompassing co-authorship[a]
Yes	Yes	38	up to 20%
No	Yes	11	up to 5%
No	No	6	none
Yes	No	1	none

[a] All members of the group are co-authors of a joint publication at least once (see Parthey, H., 1996, Fig. 2)

Table 7.2 Percentage of scientists working interdisciplinarily and group composition (education in different disciplines) in 41 research groups from three non-university institutes of the life sciences at the beginning of the 1980s (each"+" represents one group)

Group composition (education in different disciplines)										
%	Z = 0.0	Z = 0.1	Z = 0.2	Z = 0.3	Z = 0.4	Z = 0.5	Z = 0.6	Z = 0.7	Z = 0.8	Z = 0.9	Z = 1.0
100				+				++	+	+	
90								++			
80	+					++			+		
70	+					+		++			
60	+				++	+					
50	++										
40	++				+	+++			+		
30	++			+					+		
20	+							+	+		+
10					+						
0	+++			+							

Note. Z = 0 if only one discipline is represented in the group; Z = 1 if different disciplines are equally represented, in the analysis for a maximum of six disciplines: mathematics, physics, chemistry, biology, agricultural science, and medicine.

However, the finding in Table 7.3 underlines that *only practiced interdisciplinarity of problem and method (4) significantly correlates with co-authorship (6)*. We see concurrent rankings, i.e., the more/less that individual scientists within the group practice interdisciplinarity of problem and method, then the more co-authorship within the group increases/decreases. This corresponds to the finding we see in Table 7.1: encompassing co-authorship can only be expected in the case of interdisciplinarity of both problem and method.

Studies on personal interdisciplinarity in science touch on the analysis of collaborative work in research groups, especially regarding the *influence of other group members on the performance of an interdisciplinarily working scientist*. In good tradition, social science questions the influence of others on one's own performance or the advantages/disadvantages of working in groups compared to individual work

Table 7.3 Interdisciplinarity and co-authorship

	(2) Distribution of expertise by discipline	(3) Interdisciplinary problem formulation	(4) Interdisciplinarity of problem and method	(5) Publication rate per scientist	(6) Co-authorship in the group
(1) Multidisciplinary composition of the group	0.78	0.41	0.34	0.01	0.16
(2) Distribution of expertise by discipline	1.00	0.29	0.33	0.17	0.08
(3) Interdisciplinary problem formulation		1.00	0.29	0.19	0.26
(4) Interdisciplinarity of problem and method			1.00	0.02	0.39
(5) Publication rate per scientist				1.00	0.00

Underlined coefficients are significant with at least 5% probability of error (Spearman's rank correlation)

(already highlighted by Triplett in 1898). This question, applied to the scientific work itself, leads to analysis of the relationship between individual and cooperative performance in research groups and follows Max Planck's view of science, as mentioned previously, that their separation into different subjects "is not based on the nature of the matter, but only on the limitations of human capacity, which inevitably leads to a division of labor" (Planck, 1944, p. 243, translated).

In this way, the analysis of research groups deals with a topic of long-standing interest to scientists from different disciplines—and especially science researchers—and that remains highly topical today. Research of this kind has existed worldwide since the 1930s. Such research tends to be based on different methods such as participating observation or historical reconstruction. The more or less standardized questioning of research groups only began in the 1960s. In particular, the assumptions and methods used in the 1960s and 1970s assume that the effectiveness of research groups is decisively influenced by the *correspondence between the structure of the problem and the division of labor within the research group* (e.g., Pelz & Andrews, 1966).

These studies examined the working relationships that researchers must enter into with each other when working on certain problem areas. By *problem structure* we mean, above all, the relationships between the primary, secondary, and subtopics of a problem field. On the basis of numerous analyses from the 1960s and

1970s, the concept of a research group has developed that can be characterized by the following features (e.g., Swantes, 1970):

- common concern, in the form of a problem field to be worked on together,
- division of labor and cooperation in methodological problem solving, and
- their coordination by leadership.

However, empirical tests reveal that the above-mentioned assumption (that the effectiveness of research groups is decisively influenced by the correspondence between the problem structure and the structure of the division of labor in the group) is only supported to a limited extent. Thus, our analyses—which are also recorded in larger overview studies on German science research (Woodward, 1985)—point to two fundamental considerations: On the one hand, the existence of a problem situation and correspondingly formulated research problems are certainly necessary for the development of cooperative relationships between researchers, but they are not sufficient. *The necessary and sufficient condition for forms of cooperation between scientists to occur is the existence of a research situation with regard to a problem, i.e., above all, the creation and actual availability of conceptual and physical means to deal with the problem.* On the other hand, different types of research situations exert different influences on the form of cooperation. Different degrees of availability of idealistic and material means for the treatment of research problems require different relations between researchers based on the division of labor.

Our empirical finding is that it is not the composition of a group of representatives from different scientific disciplines that is significantly correlated with co-authorship, *but only the group share of scientists who practice interdisciplinarity of problem and method.* According to our analyses, the decisive characteristic of interdisciplinary research situations is therefore not—as was often assumed in the first approach of sociological studies into interdisciplinarity—the multidisciplinary composition of the group according to education and competence in different disciplines, but rather *the disciplinary lack of knowledge about problem solving among individual scientists and the resulting search for method transfer from other special fields.*

Institutionalization of Research Situations

Research institutes were and are designed as self-organizing systems. It is also a goal to develop research–technical systems that have self-organizing properties in science. Self-organizing systems are constantly confronted with alternatives in which it is up to them to make a selection. In this sense, researchers are always in situations where they have to decide for or against performing certain actions. Description and explanation of scientific institutions can be based on the fact that there is a fundamental need for a social space for the creation and development of research situations, without which science cannot exist, as its history shows.

The researcher needs the institution, because this is the only way to ensure the necessary *freedom* for research. This freedom is created through appropriate funds, such as the personnel and material budgets, and through the institute's own system of information, communication, and library. *Scientific libraries* as a component of scientific institutions become scientific workplaces to the extent that they make their publications available for further research with minimal redundancy as the sciences become increasingly differentiated. And the researchers themselves decide on the necessary and sufficient minimization of redundancy. The exchange of letters between scholars has shown—and continues to show—this in an exemplary manner at its time. Today, *scientific journals* have taken over the function as "libraries of scientific disciplines" (Parthey, 2003). Here, researchers are responsible in the function of editors, on behalf of scientific collectives (i.e., the networks of journals), of which at least two scientists (i.e., peers) judge the submitted research findings of others (i.e., peer review) according to whether (and after which revisions) they should be included in the respective special library of a "scientific journal." Publications in scientific journals contain, at least in one structural part, something scientifically new, which is presented along with exact citations that form a comprehensible reference to the "old" in science. Since its emergence in the second half of the seventeenth century, the scientific journal has proved its worth as an organ in the communication and information system of original research papers.

Scientific disciplines differ according to the area of investigation of reality and the theory on which it is based; how further knowledge of the structure and laws of the world is sought; which of the problems and which methodological approaches are preferred for their scientific treatment. *Disciplinarity* in science can be increasingly differentiated. The reasons for this are the increasingly higher degree of specialization of this knowledge and the discipline specific terminology created for its articulation, as well as the highly specialized research techniques required to further deepen this specialized knowledge. In this sense, it can be observed that new scientific disciplines have emerged at universities to the extent that, first, a chair has been created for each new scientific discipline; second, a textbook has been written for it; and, third, after the advent of letterpress printing, a new journal has become available for original papers by researchers in this new scientific discipline. Umstätter (2003) points to a comparatively "constant relation of journals and special fields." Wilhelm Ostwald has described this process of organizing new journals (in the process of the development of a new scientific discipline, which he himself helped to promote) as follows:

> That I then, after the textbook was finished, founded the *Zeitschrift für Physikalische Chemie* (Journal of Physical Chemistry) was just as natural a process... The fact that both forms of organizational work, the textbook and the journal, had a considerable influence on the further development of matters, is essentially due to the fact that at that time (in the eighties of the last century) a number of excellent collaborators in the field appeared at far-flung points in the cultural world, i.e., without mutual agreement or influence, who very soon made the scientific content of the field unusually rich and fruitful. They found the ground prepared by the above-mentioned works, and conversely, the new journal was able to prove its raison d'être to other circles by publishing groundbreaking works soon. (Ostwald, 1919, p. 10, translated)

Although the emergence of institutions is generally explained in terms of people's demand for individual orientation and social order, considerations in institutional theory also point out that institutions are only accepted and supported in people's demand for individual orientation and social order to the extent that they do not conflict with their interests. In this sense, forms of scientific institutions in their historical formation are of particular interest.

Securing Science and Its Freedom Through Institutions: Historical Forms of Institutionalization

As outlined above, the scientifically active person needs the institution, because only through this can the free space necessary for research be secured. This free space is created by appropriate funds, such as personnel and material budgets, and with an institute's own system of information, communication, and library. In order to be attractive, the scientific institution must secure the researcher an appropriate status in society and itself be flexible enough to cope with the dynamics of the modern scientific enterprise.

Plato's Academy Near Athens, Aristotle's High School in Athens and State Research Center in Alexandria

Obviously, the history of scientific institutions begins with the fact that Plato gathered his students around him in a grove of the Academy near Athens since about 388 B.C.E. Thus, the Platonic Academy was also the first scientific institution. Aristotle was active in this academy for 19 years until Plato's death. Afterwards, he was appointed by the Macedonian King Philip II as tutor to his son Alexander. Soon after Philip's death, Aristotle returned to Athens and founded his own school, the Lyceum, as a second scientific institution for teaching young people.

As the third scientific institution, a state study center of the entire Hellenistic world was established in Alexandria in the third century B.C.E., consisting of the Mouseion research center (cf. Parthey, G., 1838) and the largest library of the ancient world. Euclid was among those who worked there, between 320 and 260 B.C.E. and Ptolemy from 127 to 141 B.C.E., who carried out the observations used in his work "Almagest" in the observatory. Alexandria was a center of scientific life for more than 700 years of history until about the beginning of the fifth century C.E. In the following centuries without any scientific institutions worth mentioning, hardly any scientific publications were published, sometimes not at all, i.e., for several centuries almost no scientists can be proven.[1]

---

[1] See Grau (1988); Parson & Platt (1973); Rüegg (1993, 1996, 2004); Mieg & Evetts (2018).

University Education of Science-Based Professions Since the Middle Ages

Even if the institutions that emerged in antiquity to ensure problematization and methodical problem-solving—such as the Platonic Academy, the Aristotelian Lyceum as a municipal grammar school, and the Alexandrian Mouseion as a state research institution—did not survive the centuries despite their research achievements, a new, sustainable scientific institution has endured since the twelfth century with the university, due to the increasing interest in the training of science-based professions (initially mainly for doctors and lawyers). From then on, the university has also been involved in the training of other emerging science-based professions and has thus become a fundamental institution of science all over the world. In addition to this, modern academies have also been established with worldwide success since the fifteenth century (following the Platonic Academy) as research institutions without the teaching obligations of universities.

Today, university education can enable people to carry out a scientific activity if, in addition to imparting a disciplinary field of knowledge that is subject to constant renewal, it aims above all at the ability to independently ask further questions, to develop these into knowledge problems with the available level of knowledge and to methodically gain problem-solving insights. This can only be achieved by teaching that presents and discusses the process of scientific knowledge in a model way and actively involves the students in this process. Research-based learning is thus an integral part of every scientific course of study (cf. Mieg et al., 2021).

Non-university Research Institutes Since the Emergence of Science-Based Economy

In the nineteenth century, the institutional form of science was still largely that of the academy and—increasingly—the university in the unity of teaching and research striven for by Wilhelm von Humboldt, whereby his great scientific plan called for independent research institutes as integrating parts of the overall scientific organism in addition to the Academy of Sciences and the university (cf. Humboldt, 1964). With the emergence of science-based industries such as the electrical industry, which could not have existed beforehand—not even as a trade—without the scientific theories of flowing electricity and electromagnetism and the discovery of the dynamoelectric principle (1866 by Werner von Siemens), and the transformation of traditional trades into science-based industries such as the chemical industry in the last third of the nineteenth century (cf. Zott, 1998), the university was able to establish itself as a center of scientific excellence. In the last third of the nineteenth century, the establishment of scientific institutions outside of universities grew to include large chemical research laboratories set up by the chemical industry, and state laboratories for basic research in physics that were intended to contribute to improving the scientific basis of precision measurement and materials testing. An example of the latter is the Physikalisch-Technische Reichsanstalt, founded in 1887 in Berlin-Charlottenburg (Imperial Physical-Technical Institute, cf. Förster, 1887; Cahan, 1989), which Wilhelm Ostwald still described as a "completely new type of scientific institution" two decades later (Ostwald, 1909, p. 294).

The Physikalisch-Technische Reichsanstalt consisted of two departments, the scientific and the technical. The first one is currently still trying to work on problems of physical precision measurement that are pending but urgently in need of a solution, especially those problems for which universities lack the necessary rooms and equipment or those that require a scientist to devote themselves entirely to research for a long period of time without the additional demands of teaching. The second department is intended to provide direct support to the precision trade, taking care of all the technical services that cannot be carried out by mechanics in small- and medium-sized enterprises, but also serving as an official testing institute for mechanical and technical instruments. The president of the institute is also the director of the scientific department.[2] The success of the Physikalisch-Technische Reichsanstalt triggered efforts to establish an analogous Chemisch-Technische Reichsanstalt (Imperial Chemical-Technical Institute). Driven by the developmental needs of science itself as well as of the state and the economy, which is also evident in studies of science policy in Germany since the eighteenth century (cf. McClelland, 1980), several research institutes independent of teaching were founded in Berlin within the framework of the Kaiser-Wilhelm-Gesellschaft zur Förderung der Wissenschaften (KWG, Kaiser Wilhelm Society for the Promotion of Science), which existed for more than three decades (1911–1945) and was financed by both the state and the economy. Today, the former Kaiser-Wilhelm-Gesellschaft (KWG) has been succeeded by the Max-Planck-Gesellschaft (Max Planck Society, MPG, www.mpg.de).

Interdisciplinary Research Situation in Non-University Research Institutes: Lessons from the Example of the Kaiser-Wilhelm-Gesellschaft (KWG)

As early as the last third of the nineteenth century, research directions developed "that no longer fit into the university framework at all, partly because they require such large mechanical and instrumental facilities that no university institute can afford them, and partly because they deal with problems that are far too advanced for students and can only be presented by young scholars." (MPG, 1961, p. 82, translated) It also addresses novel relationships between research in government institutes and in the business world. For example, Adolf von Harnack, in his

---

[2] The first president of the Physikalisch-Technische Reichsanstalt was Hermann von Helmholtz. For the beginnings of science promotion by science-based economy see Kant (2002).

memorandum of November 1909, used as an example the situation in organic chemistry, "the leadership of which until not so long ago lay undisputedly in the chemical laboratories of German universities," but which "today has almost completely migrated from there to the large laboratories of factories," and concluded that "this whole field of research is to a large extent lost to pure science," because "factories always continue research only to the extent that it promises practical results, and they keep these results as secrets or put them under patent. Therefore, the laboratories of the individual factories, which work with the greatest of means, can rarely be expected to promote science. The reverse has always been true: pure science has brought the greatest support to industry by opening up truly new areas." (MPG, 1961, pp. 82–83, translated)

Thus, with the emergence of research-dependent industries, such as the chemical and electrical industries in the last third of the nineteenth century, there was an increase in the founding of scientific institutions outside the universities, for example, large chemical research laboratories, which the chemical industry built up, in addition to state laboratories for physical research that were to contribute to the improvement of the scientific basis of precision measurement and materials testing. There are three main reasons given for the establishment of research institutes that are independent of teaching (financed not only by the state but also by industry):

1. First, the rising costs of research technology (cf. Biedermann, 2002).
2. Second, the growing teaching obligations for university teachers, which make it difficult for them to work in the unity of teaching and research that Wilhelm von Humboldt aimed for.
3. And thirdly, the opportunities to create and work on many more interdisciplinary research situations, unhindered by the inevitably disciplinary teaching profiles at universities.

Therefore, in the founding history of the Kaiser-Wilhelm-Gesellschaft (KWG), reference was made to the fruitfulness of collaboration between researchers from different directions. From a later viewpoint of Adolf Butenandt, the founding of the KWG took place in 1911 in order to:

...close a gap in the German scientific structure. It was felt that working methods became necessary that were difficult to master in the conventional forms: It seemed urgently necessary to allow scholars who wanted to devote themselves primarily to pure research to work in complete freedom, to shield them to a large extent from all those things that might ultimately impair their ability to perform in the service of human progress. Secondly, it was necessary to give scholars working in newly developing border regions their very special working instrument, tailored to their needs, in order to strengthen and grow disciplines which had no or not yet had sufficient space in the structure of the universities and technical colleges. From the early days of the Kaiser Wilhelm Society, I mention as examples the physical chemistry of Haber, the radiochemistry of Hahn, the theoretical physics of Einstein, the biochemistry of Warburg. Thirdly, since the founding of the Kaiser-Wilhelm-Gesellschaft, the task had existed of developing and supervising new types of institutes. In order to solve some of the problems, very extensive personnel and material resources must be combined to form a structure that would have to go beyond the scope and technical complexity of any university structure. The institutes for iron research, coal research, and occupational physiology can be mentioned as examples. (MPG, 1961, pp. 7–8, translated)

August von Wassermann demanded at the inauguration of the Kaiser Wilhelm Institute for Experimental Therapy (as an institute of the KWG) in October 1913, that:

> New ways of healing and all that is connected with it, especially the recognition of disease, should no longer be left here in this house to the more or less subjective experiences of the individual observer at the sickbed, as in earlier times, but should be explored on the basis of purposeful research with the help of the exact scientific auxiliary disciplines. (MPG, 1961, p. 158, translated)

Thus, in the founding history of the KWG, the fertility of a traffic of researchers from different directions was pointed out. Especially in the justifications for life science research without additional responsibility for teaching, the idea was developed that they should work outside the university in a more interdisciplinary way, which has also been scientifically profitable (cf. Jaekel, 1907). To this end, institutes for biochemistry and biophysics were founded in the KWG, among others. In most cases, a successfully proven "horizontal" interdisciplinarity led to the development of new disciplines with all the characteristics of an independent discipline, including later university teaching and training institutes.

Interdisciplinarity as a developmental form of science, which is institutionalized in a disciplined manner in the further scientific procedure, including within the university framework, is more of a "horizontal" interdisciplinarity, less of a "vertical" one,[3] as it was pursued outside of the university and instead within the framework of the KWG, especially in the institutes devoted to brain research, iron, metals, coal, leather, hydrobiology, silicates, fluid dynamics, and plant breeding programs. A comparison of the development of science in the USA and in Germany over the past decades shows that *vertical interdisciplinarity* is institutionalized and evaluated in non-university research institutions in the USA more quickly than elsewhere and, if scientific success continues, is also more rapidly introduced to university education programs.

A further lesson relates to the issue of research funding. In recent decades, the design of research situations has led to considerations that, in terms of their institutionalization, large-scale research should be set up in the form of umbrella organizations and so-called *virtual research institutes*. In Germany, for example, a new Joint

---

[3] Parthey and Schreiber (1983) introduced the distinction between horizontal vs. vertical interdisciplinarity as follows: "The importance of science is based not only on its new fundamental findings, but increasingly on the profound social impact of its results. For science to become fully effective, it is usually necessary to transfer the principles of basic research into social, especially economic, practice. This generally requires, as in the case of the development of a new drug or plant protection product, closed lines of cooperation from basic research through industrial development and production of active ingredients to application research and preparation for use of the new preparations in health care or agriculture, which include interdisciplinary or at least multidisciplinary cooperation between representatives of very different disciplines. We would like to distinguish such "vertical" levels of interdisciplinarity oriented towards a transfer goal from the so-called "horizontal" interdisciplinarity, which is to be practiced between representatives or collectives of different disciplines of basic research and is almost exclusively the focus of this publication." (pp. 304–305, translated)

Science Conference (Gemeinsame Wissenschaftskonferenz, GWK, www.gwk--bonn.de) was set up in 2008, which deals with the funding of the major research organizations such as the German Research Foundation (Deutsche Forschungsgemeinschaft, DFG), the Max Planck Society (Max-Planck-Gesellschaft, MPG), the Helmholtz Association of German Research Centres (Helmholtz-Gemeinschaft Deutscher Forschungszentren) and the Leibniz Association (Leibniz Gemeinschaft), as well as academies such as the Leopoldina Academy of Sciences and the Wissenschaftskolleg in Berlin. This German science conference funds research projects, research buildings, and large-scale research equipment at universities that are of supraregional importance.

For a long time now, around two-thirds of all investment in research and development in Germany has come from the industrial sector. In order to tie in with this innovation momentum, the Helmholtz Association (www.helmholtz.de), Germany's largest non-university scientific organization, is about to integrate the entire chain of effects from basic and applied research to future product maturity (*vertical* interdisciplinarity). In doing so, it too is relying on a strategic partnership with the universities. The fifteen Helmholtz centers are involved in areas of special university research and in priority programs funded by the German Research Foundation.

The discussion on science and financial policy in Germany since the beginning of the twentieth century shows that with the enabling financing of science by the innovative power of the economy, there is also a change in research in a science-integrated economy (cf. Spur, 2002), which may not solidify every new field of knowledge into a teachable discipline. In this context, we should to refer to the *methodological structure of research situations as an invariant of knowledge production* even in the twenty-first century but now with more focus on interdisciplinary research situations and their institutionalization than in the preceding centuries.

Ambivalence of the Experimental Method in Research

Science as published methodical problem solving has today for this purpose three large methodological structures: the experimental, the mathematical, and the historical method. At the birth of science, mainly the bare observation method, the mathematical method, and the historical method were used, because there was such a strict distinction between the epistemological and the technological that experimental methods for revealing truths were rejected in preference for only the bare observation without experiment.

Experimentation was excluded at the birth of science, due to the argument of ensuring scientific integrity in the methodological procedure of research. This remained true for science for one and a half millennia. Only with Galileo Galilei was there adoption of experimentally based observation—in all those cases where the truth value of statements cannot be determined directly by bare observation. For Galileo, this experimental basis for observation allowed more robust investigation and confirmation of hypothesized connections and claims of fact. Identifying an experimental problem, carrying out experiments, and finally interpreting experimental results for the verification of *hypotheses* were also introduced into research with Galileo, as three steps in the experimental method. In research, experiments are characterized by a system of conditions deliberately set by the experimenter so that essential relationships can be observed in a repeatable way under conditions of change and control.

Unlike mere observation, experimentation is based on an *active intervention* in natural and social contexts in the form of experimental technology, the ambivalence of which has now been discussed more intensively again since the twentieth century in various studies, following Aristotle's rejection of experimentally conditioned observation in research.

Ambivalence, following its psychological usage, is used to describe an often-conflictual state in which opposing courses of action, such as affection-rejection, exist simultaneously with respect to the same object (see Bleuler, 1914/2017). Experimental research is increasingly ambivalent about its impact on society and science. A historically early example of the ambivalence of experimental research in the twentieth century is Lise Meitner's rejection in July 1938 of Fritz Straßmann's first laboratory notes on nuclear fission in uranium irradiated with neutrons (by chemical detection of barium in the irradiation products).[4] When Straßmann and Otto Hahn turned to this experiment again in December 1938 and had to conclude on uranium nuclear fission, they first communicated this to Meitner, who had emigrated in the meantime, with their publication submitted for printing (Hahn & Straßmann, 1939). Within a few days, Meitner had calculated the energy balance of this nuclear fission process in an article published jointly with Otto Frisch in January 1939 (Meitner & Frisch, 1939). Concerning the ambivalence of science, Karl Friedrich von Weizsäcker concluded:

> Science cannot afford, under the motto that it seeks the truth and nothing else, not to consider the effects it has on life. Personally, I have never found it comprehensible that scientists felt that when what science produces in technology is used by politicians or by the military in such a way that scientists are unhappy with it, to say that here science has been misused. After all, science has provided these means, and it is, of course, responsible for the means it puts into other hands. If it supplies means into a political structure that is not adequate to these means—means that have a baleful effect in this structure—the least that is to be demanded of science is that it reflect on how the structure can be changed, which obviously cannot avoid producing these baleful effects. In this sense, then, self-reflection of science is a demand on science. (Weizsäcker, 1970, translated).

In resonance with the discussion about the scientific ambivalence of nuclear power, the German government pushed through a resolution (as early as 2000) to gradually shut down nuclear power plants. Finally, the disaster in Fukushima, Japan, led to the final legal nuclear phase-out in Germany, adopted in 2010. A new law on the nuclear phase-out in Germany is emerging for disposal of the radioactive legacies of nuclear power plants. Under the new law, the state will assume financial and organizational responsibility for nuclear waste disposal. Nevertheless, the revised draft law states

---

[4] Probably for a mix of ethical and scientific reasons. See also Krafft (1981).

that the polluter pays principle will be strictly adhered to in disposing of nuclear waste. Accordingly, energy utility companies will remain responsible for decommissioning and demolishing the nuclear power plants that they operated. For the disposal of nuclear waste, they are to pay a total of around 23.4 billion Euros into a public fund from existing reserves plus an additional risk surcharge.

In our century, *embryo research*, in particular, is increasingly ambivalent in its impact on society and science.[5] The German Embryo Protection Act prohibits the production or use of embryos for any purpose other than to induce pregnancy. Experimentation on human embryos is still a criminal offense in Germany.[6] Until now, active intervention in the human genome has also been ethically taboo internationally.

It is therefore not surprising that some researchers now fear that they have opened a Pandora's box with the CRISPR gene-editing process. A technique that is suitable for transforming yeast cells, mice, or monkeys is also suitable for creating custom-made humans. Ethicists and lawyers have spoken out internationally, and of course the luminaries of the CRISPR guild: Jennifer Doudna, the method's discoverer, and her colleague Emmanuelle Charpentier, who has moved to the Max-Planck-Institut für Infektionsbiologie (Max Planck Institute for Infection Biology) in Berlin. In order to equip mice with enhanced cancer protection, cell biologists introduced a mutation into their genome that activates a tumor-suppression gene (Morris et al., 2012). Only afterwards were they surprised to discover that the genetically manipulated animals showed side effects of premature ageing.

Such unexpected effects constitute perhaps the most cogent argument opposing human modification of our genome. The approximately twenty thousand human genes are interwoven into an immeasurably complex network of reciprocal influences. Any intervention will have consequences, and by no means all of them can be predicted. All over the world, biotechnologists and ethicists are discussing whether there can be circumstances in which it is acceptable to impose such potential consequences on future generations.

References

Andrews, F. M. (Ed.). (1979). *Scientific productivity: The effectiveness of research groups in six countries*. Cambridge University Press/UNESCO.

Bayertz, K. (2003). Die Wahrheit über den moralischen Status menschlicher Embryonen. In: G. Mario u. H. Just (Eds.). Die Forschung an embryonalen Stammzellen in ethischer und rechtlicher Perspektive (pp. 178–195) Nomos.

Biedermann, W. (2002). Zur Finanzierung der Institute der Kaiser-Wilhelm-Gesellschaft zur Förderung der Wissenschaften Mitte der 20er bis Mitte der 40er Jahre des 20. Jahrhundert. In

---

[5] See Habermas (2001); Bayertz (2003).

[6] Gesetz zum Schutz von Embryonen vom 13. Dezember 1990 (Embryonenschutzgesetz – ESchG). Bundesgesetzblatt I, 2746.

H. Parthey & G. Spur (Eds.), *Wissenschaft und Innovation: Jahrbuch Wissenschaftsforschung 2001* (pp. 143–172). Gesellschaft für Wissenschaftsforschung.

Bleuler, E. (1914). Die Ambivalenz. Festgabe zur Einweihung der Neubauten der Universität Zürich 18. IV. 1914 (Festgabe der medizinischen Fakultät, pp. 95–106). Zürich: Schulthess & Co. [Reprinted in K. Fischer & H. Parthey (Eds.) (2019). Ambivalenz in der Wissenschaft: Wissenschaftsforschung Jahrbuch 2017. WBV. Open access]

Cahan, D. (1989). *An institute for an empire. The Physikalisch-Technische Reichsanstalt 1871–1918.* Cambridge University Press.

Förster, W. (1887). Die Physikalisch-Technische Reichsanstalt. Berlin 1887 (Broschure)

Grau, C. (1988). *Berühmte Wissenschaftsakademien: Von ihrem Entstehen und ihrem weltweiten Erfolg.* Harry Deutsch.

Habermas, J. (2001). *Die Zukunft der menschlichen Natur. Auf dem Weg zu einer liberalen Eugenik?* Suhrkamp.

Hahn, O., & Straßmann, F. (1939). Über das Zerplatzen des Urankerns durch langsame Neutronen. In: Abhandlungen der Preußischen Akademie der Wissenschaften (Berlin), 12.

Humboldt, W. Von, (1964). Über die innere und äußere Organisation der höheren wissenschaftlichen Anstalten in Berlin. In: Humboldt, W. Von., Werke in fünf Bänden. Band IV, Schriften zur Politik und zum Bildungswesen (pp. 255–266). Akademie-Verlag.

Kant, H. (2002). Aus den Anfängen der Wissenschaftsförderung durch wissenschaftsbasierte Wirtschaft: Herrmann Helmholtz, Werner Siemens und andere. In H. Parthey & G. Spur (Eds.), *Wissenschaft und Innovation: Jahrbuch Wissenschaftsforschung 2001* (pp. 129–142). Gesellschaft für Wissenschaftsforschung.

Krafft, F. (1981). *Im Schatten der Sensation: Leben und Wirken von Fritz Strassmann.* Verlag Chemie.

Jaekel, O. (1907). Über die Pflege der Wissenschaft im Reich. In: Der Morgen, 20, 617–621.

Meitner, L., & Frisch, O. R. (1939). Disintegration of uranium by neutron: A new type of nuclear reaction. *Nature, 143,* 239–240.

MPG Max-Planck-Gesellschaft zur Förderung der Wissenschaften, Generalverwaltung. (1961). 50 Jahre Kaiser-Wilhelm-Gesellschaft und Max-Planck-Gesellschaft zur Förderung der Wissenschaften 1911–1961. Beiträge und Dokumente. Hubert & Co.

McClelland, C. E. (1980). *State, society and University in Germany 1700–1914.* Cambridge University Press.

Mieg, H. A., Ambos, E., Brew, A., Galli, D., & Lehmann, J. (Eds.). (2021). *The Cambridge handbook of undergraduate research.* Cambridge University Press.

Mieg, H. A., & Evetts, J. (2018). Professionalism, science, and expert roles: A social perspective. In K. A. Ericsson, R. R. Hoffman, A. Kozbelt, & A. M. Williams (Eds.), *The Cambridge handbook of expertise and expert performance* (2nd ed., pp. 127–148). Cambridge University Press.

Morris, S. A., Grewal, S., Barrios, F., Patankar, S. N., Strauss, B., Buttery, L., Alexander, M., Shakesheff, K. M., & Zernicka-Goetz, M. (2012). Dynamics of anterior-posterior axis formation in the developing mouse embryo. *Nature Communication, 3,* 673.

Ostwald, W. (1909). *Große Männer.* Akademische Verlagsgesellschaft.

Ostwald, W. (1919). Handbuch der allgemeinen Chemie. Band I: Die chemische Literatur und die Organisation der Wissenschaft. Leipzig: 1919. S. 10. 43

Parsons, T., & Platt, G. M. (1973). *The American University.* Harvard University Press.

Parthey, G. (1838). *Das Alexandrinische Museum.* Nicolaische Buchhandlung.

Parthey, H. (1982). Wissenschaftsmetrische Analyse der Verteilung von Autoren nach Publikationsraten und Wissenschaftsdisziplinen in biowissenschaftlichen Forschungsinstituten der siebziger Jahre des 20. Jahrhunderts. In: H. Parthey, D. Schulze, A. A. Starcenko & I. S. Timofeev (Eds.), Methodologische Probleme der Wissenschaftsforschung, Teil Wissenschaftsmetrische Methoden. Wissenschaftswissenschaftliche Beiträge, 17, 1–16.

Parthey, H. (1983). Forschungssituation interdisziplinärer Arbeit in Forschergruppen. In H. Parthey & K. Schreiber (Eds.), *Interdisziplinarität in der Forschung: Analysen und Fallstudien.* Akademie-Verlag.

Parthey, H. (1990). Relationship of interdisciplinarity to cooperative behavior. In P. H. Birnbaum-More et al. (Eds.), *International research management* (pp. 141–145). Oxford University Press.

Parthey, H. (1996). Kriterien und Indikatoren interdisziplinären Arbeitens. In P. W. Balsiger, R. Defila, & A. Di Giulio (Eds.), *Ökologie und Interdisziplinarität – Eine Beziehung mit Zukunft* (pp. 99–112). Birkhäuser.

Parthey, H. (1997). Analyse von Forschergruppen. In H. Bertram (Ed.), *Soziologie und Soziologen im Übergang: Beiträge zur Transformation der außeruniversitären soziologischen Forschung in Ostdeutschland* (pp. 543–559). Leske + Budrich.

Parthey, H. (2003). Zeitschrift und Bibliothek im elektronischen Publikationssystem der Wissenschaft. In H. Parthey & W. Umstätter (Eds.), *Wissenschaftliche Zeitschrift und Digitale Bibliothek: Wissenschaftsforschung Jahrbuch 2002* (pp. 9–46). Gesellschaft für Wissenschaftsforschung.

Parthey, H. (2011). Institutionalisierung disziplinärer und interdisziplinärer Forschungssituation. In K. Fischer, H. Laitko, & H. Parthey (Eds.), *Interdisziplinarität und Institutionalisierung der Wissenschaft: Wissenschaftsforschung Jahrbuch 2010* (pp. 9–35). WVB Wissenschaftlicher Verlag.

Parthey, H. (2019). Ambivalenz der experimentellen Methode in der Forschung. In K. Fischer & H. Parthey (Eds.), *Ambivalenz in der Wissenschaft: Wissenschaftsforschung Jahrbuch 2017* (pp. 57–60). Wissenschaftlicher Verlag Berlin.

Parthey, H. (2020). Sicherung der Wissenschaft durch Institutionen in der Antike, im Mittelalter und in der Neuzeit. In: H. A. Mieg, H. Lenk & H. Parthey (Hrsg.). (2020). Wissenschaftsverantwortung: Wissenschaftsforschung Jahrbuch 2019 (pp. 153–164). : wvb Wissenschaftlicher Verlag Berlin.

Parthey, H., & Schreiber, K. (1983). Voraussetzungen und Formen interdisziplinärer Forschung. In H. Parthey & K. Schreiber (Eds.), *Interdisziplinarität in der Forschung: Analysen und Fallstudien* (pp. 303–309). Akademie-Verlag.

Pelz, D. C., & Andrews, F. M. (1966). *Scientists in organizations: Productive climates for research and development*. Wiley.

Planck, M. (1944). Ursprung und Auswirkungen wissenschaftlicher Ideen (Lecture given on February 17, 1933, at the Association of German Engineers, Berlin). In M. Planck (Ed.), *Wege zur physikalischen Erkenntnis; Reden und Vorträge*. S. Hirzel.

Rüegg, W. (Ed.). (1993). *Geschichte der Universität in Europa. Band I: Mittelalter. Hrsg. v. Walter Rüegg*. C. H. Beck.

Rüegg, W. (Ed.). (1996). *Geschichte der Universität in Europa. Band II: Von der Reformation bis zur Französischen Revolution (1500–1800). Hrsg. v. Walter Rüegg*. C. H. Beck.

Rüegg, W. (Hrsg.). (2004). Geschichte der Universität in Europa, Band III: Vom 19. Jahrhundert zum Zweiten Weltkrieg (1800–1945). Edited by Walter Rüegg. C. H. Beck.

Spur, G. (2002). Wandel der Forschung in einer wissenschaftsintegrierten Wirtschaft. In H. Parthey & G. Spur (Eds.), *Wissenschaft und Innovation: Wissenschaftsforschung Jahrbuch 2001* (pp. 41–57). Gesellschaft für Wissenschaftsforschung.

Steck, R. (1979). Organisationsformen und Kooperationsverhalten interdisziplinärer Forschergruppen im internationalen Vergleich. In F. R. Pfetsch (Ed.), *Internationale Dimensionen in der Wissenschaft* (pp. 87–108). Institut für Gesellschaft und Wissenschaft an der Universität Erlangen-Nürnberg.

Stegmüller, W. (1983). Vom dritten bis sechsten (siebten?) Dogma des Empirismus. In P. Weingartner & J. Czermak (Eds.), *Erkenntnis-und Wissenschaftstheorie Akten des 7. Internationalen Wittgenstein Symposiums* (pp. 232–244). Hölder-Pichler-Tempsky.

Stegmüller, W. (1986). *Probleme und Resultate der Wissenschaftstheorie und Analytischen Philosophie. Band II: Theorie und Erfahrung. Dritter Teilband: Die Entwicklung des neuesten Strukturalismus seit 1973*. Springer.

Swantes, G. M. (1970). The social organization of a university laboratory. *Minerva., 8*(1), 36–58.

Triplett, N. (1898). The dynamogenic factors in pacemaking and competition. *American Journal of Psychology, 9*, 507–532.

Umstätter, W. (2003). Was ist und was kann eine wissenschaftliche Zeitschrift heute und mor-
    gen leisten. In: Wissenschaftliche Zeitschrift und Digitale Bibliothek: Wissenschaftsforschung
    Jahrbuch 2002. Hrstg. v. Heinrich Parthey u. Walther Umstätter (pp. 143–166). : Gesellschaft
    für Wissenschaftsforschung.
Weizsäcker, K. F. V. (1970, 13 July). Die Macht der öffentlichen Meinung im Kampf gegen
    Einzelinteressen. In: Süddeutsche Zeitung, 166, p 7.
Woodward, W. R. (1985). Committed history and philosophy of the social science in the two
    Germanies. *History of Science., 23*(1), 25–72.
Zott, R. (1998). Die Umwandlung traditioneller Gewerbe in wissenschaftsbasierte Industriezweige:
    das Beispiel chemische Industrie—das Beispiel Schering. In S. Greif, H. Laitko, & H. Parthey
    (Eds.), *Wissenschaftsforschung: Jahrbuch 1996/97* (pp. 77–95). BdWi-Verlag.

Chapter 8
The Application of the Precautionary Principle in the EU

Kristel De Smedt and Ellen Vos

Abstract The precautionary principle is a guiding principle that allows decision makers to adopt precautionary measures even when scientific uncertainties about environmental and health impacts of new technologies or products remain. It is also a debated principle. Proponents of the precautionary principle argue that it provides a framework for improving the quality and reliability of decisions over technology, science, ecological and human health, and leads to improved regulation. Opponents argue that it is incoherent, lacking orientation and that it hinders innovation. The aim of this Chapter is to increase understanding of the perceived tension between the precautionary principle and innovation by examining how the precautionary principle is applied in EU law and by the EU courts. This Chapter is based on the findings of an EU-funded research project entitled REconciling sCience, Innovation and Precaution through the Engagement of Stakeholders (RECIPES).

Introduction

The precautionary principle has acquired a firm place in the governance of modern society. It guides decision makers faced with risks, scientific uncertainty, and public concerns. It allows them to adopt precautionary measures even when scientific uncertainties about environmental and health impacts of new technologies or products remain. However, it is also a very debated principle, in particular with respect to its influence on innovation.[1]

---

[1] The research on which this article is based was a joint effort of all RECIPES partners. We would like to thank all partners for the feedback we received in relation to our analysis of the precautionary principle in EU law. For more information on the RECIPES project, see https://recipes-project.eu/.

K. De Smedt (✉) · E. Vos
Maastricht University, Maastricht, The Netherlands
e-mail: k.desmedt@maastrichtuniversity.nl; e.vos@maastrichtuniversity.nl

H. A. Mieg (ed.), *The Responsibility of Science*, Studies in History and
Philosophy of Science 57, https://doi.org/10.1007/978-3-030-91597-1_8

On the one hand, the precautionary principle is seen as a tool that helps scientists, innovators, policy makers, politicians, and societal organizations to reflect on which technologies need to be developed, which threshold of damage can be allowed and which level of uncertainty is acceptable to society.[2] As such, the principle functions as a theoretical construct that has created, in many different fields, a common language through which common concerns could be tackled.

As Gee argues, in the face of uncertainty, ignorance and complexity, and wider public engagement, societies could pay attention to the lessons of past experience and use the precautionary principle, to anticipate and minimize many future hazards, whilst stimulating innovation. As the case studies in the 2013 *Report on Late Lessons from Early Warnings* have shown, the timely use of the precautionary principle can often stimulate rather than hamper innovation, in part by promoting a diversity of technologies and activities, which can also help to increase the resilience of societies and ecosystems to future surprises.[3]

On the other hand, the precautionary principle has been criticized as vague, incoherent, unscientific, arbitrary and the like.[4] The precautionary principle came under attack academically especially from 2005 onwards, when Cass Sunstein portrayed it as incoherent and lacking any orientation in his 'laws of fear'.[5] Sunstein argues that risks exist on all sides of social situations, and that precautionary steps may create dangers of their own. Precautionary measures to reduce one risk may induce side-effects or so-called risk–risk trade-offs, such as increases in other countervailing risks. Therefore, Sunstein argues that in a world of risks on all sides, the precautionary principle points nowhere.[6] Moreover, critics argue that precautionary measures may be costly, and worry that measures to restrict new technologies may actually inhibit innovation.[7]

Recent times have seen the rise of 'responsible research and innovation' (RRI), indicating that innovation and precaution can go hand in hand. RRI can be defined as a "transparent, interactive process by which societal actors and innovators become mutually responsive to each other with a view on the (ethical) acceptability, sustainability and societal desirability of the innovation process and its marketable products."[8] It represents, therefore, the ongoing process of aligning research and

---

[2] Read, R., and O'Riordan, T., 'The precautionary principle under fire', *Environment: Science and Policy for Sustainable Development*, 59(5), 2017.

[3] Gee, D., 'More or less precaution', in *Late lessons from early warnings II: Science, precaution, innovation*, European Environment Agency, EEA report no 1/2013, p. 643.

[4] Zander, J., *The Application of the Precautionary Principle in Practice: Comparative Dimensions*, Cambridge University Press, New York, 2010, p. 32.

[5] Sunstein, C., *Laws of Fear: Beyond the Precautionary Principle*, Cambridge University Press, 2009.

[6] Ibid.

[7] See for instance https://laweconcenter.org/wp-content/uploads/2018/11/Portuese-Pillot-The-Case-for-an-Innovation-Principle-A-Comparative-Law-and-Economics-Analysis-2018-1.pdf.

[8] Von Schomberg, R., 'The precautionary principle: Its use within hard and soft law', *European Journal of Risk Regulation*, 2, 2012, pp. 147–156.

innovation to the values, needs and expectations of society,[9] addressing the observation that innovation—as a goal in itself—does not always lead to results that are beneficial to society as a whole or else may be accompanied by negative side effects.[10]

With this in mind, the aim of this Chapter is to increase understanding of the perceived tension between the precautionary principle and innovation, by examining how the precautionary principle is applied in EU law and by the EU courts. This Chapter is based on research carried out in the context of the EU-funded project entitled REconciling sCience, Innovation and Precaution through the Engagement of Stakeholders (RECIPES).[11] RECIPES is based precisely on this idea that the responsible application of the precautionary principle and the consideration of innovation aspects do not necessarily contradict each other. Building on this idea, the RECIPES project aims to reconcile and align science, innovation, and precaution by developing tools and guidelines to ensure that the precautionary principle is applied while still encouraging innovation. The RECIPES project is working closely with different stakeholders through interviews, workshops, and webinars.[12]

This Chapter is structured as follows. In Section "Introduction", we briefly outline how the precautionary principle is implemented in the EU. We discuss in particular the Commission Communication on the Precautionary Principle of 2000, as this document aims to give guidance on how to apply the precautionary principle at EU level.

In Section "The Precautionary Principle in EU Law", we discuss how the precautionary principle is applied in practice. First, we examine its practical application since 2000 by the EU institutions in legal acts. Taking a bird's-eye perspective, we examine whether and how the precautionary principle is explicitly applied in EU legal acts, and whether—and to what extent—the guidelines that were developed by the European Commission in its Communication have been applied. We do this by means of a literature review and an empirical study looking at all legal acts that mention the term precautionary principle.

Section "Application of the Precautionary Principle Since 2000 by the EU Institutions in Legal Acts" focuses on how the EU courts apply the precautionary principle in case law. As will be set forth below, the courts visibly struggle with this goal, and certain inconsistencies have arisen.

---

[9] Rome Declaration on Responsible Research and Innovation in Europe, 2014, https://ec.europa.eu/research/swafs/pdf/rome_declaration_RRI_final_21_November.pdf.

[10] Owen, R., Bessant, J. R., and Heintz, H., eds. *Responsible Innovation: Managing the Responsible Emergence of Science and Innovation in Society.* John Wiley & Sons, 2013.

[11] Vos, E., De Smedt, K. et al., *Taking stock as a basis for the effect of the precautionary principle since 2000*, 2020, available on https://recipes-project.eu/. We would like to acknowledge the assistance of Laura Dohmen in the carrying out of the empirical legal analysis, presented in Sections "The Precautionary Principle in EU Law" and "Application of the Precautionary Principle Since 2000 by the EU Institutions in Legal Acts".

[12] For more information, visit the RECIPES website: https://recipes-project.eu/.

Section "Application of the Precautionary Principle by the European Courts in Court Rulings from 2000 to 2019" provides some concluding remarks and will elaborate in particular on the difficulties experienced by the EU courts in reviewing precautionary measures and dealing with science and scientific uncertainty.

The Precautionary Principle in EU Law

In 1992, the Maastricht Treaty formally introduced the precautionary principle within the then EC Treaty as a principle of environmental law and policy. Article 130(2) EC Treaty (now Article 191 (2) TFEU) stipulated that the EU's environmental policy was to be based on *inter alia* the precautionary principle. It also provided that "environmental protection requirements must be integrated into the definition and implementation of other Community policies."[13]

Today the precautionary principle is generally recognized to be a principle of EU law.[14] However, Article 191(2) of the Treaty on the Functioning of the European Union (TFEU) does not provide a definition of the precautionary principle. Instead, it states:

> Union policy on the environment shall aim at a high level of protection taking into account the diversity of situations in the various regions of the Union. It shall be based on the precautionary principle and on the principles that preventive action should be taken, that environmental damage should as a priority be rectified at source and that the polluter should pay.

In order to provide guidance on the application of the precautionary principle in further EU law-making, the European Commission issued a non-binding Communication on the Precautionary Principle in 2000.[15] The Commission does not provide a definition of the precautionary principle in this Communication. Rather, the Communication sets out some constituent elements of the precautionary principle. It intends to provide guidance but not to prescribe "the final word, rather, the idea is to provide input to the ongoing debate both at Community and international level."[16] As with other general notions contained in EU legislation, the

---

[13] Morgera, E., 'Environmental law', Chapter 22 in C. Barnard and S. Peers (eds), *European Union Law*, Second Edition, Oxford University Press, 2018, p. 663.

[14] Zander, J., *The Application of the Precautionary Principle in Practice: Comparative Dimensions*, Cambridge University Press, New York, 2010, p. 81. See J. Scott on the distinction between a principle of EU law and a general principle of EU law: J. Scott, *Legal aspects of the precautionary principle: A British Academy Brexit briefing*, p. 16; see https://www.thebritishacademy.ac.uk/documents/309/Legal-Aspects-of-the-Precautionary-Principle.pdf.

[15] Communication from the Commission on the precautionary principle, COM/2000/1, hereafter COM(2000) 1.

[16] COM(2000)1, p. 8.

Communication sees European decision makers—and ultimately the courts —as responsible for elaborating the details of its application.[17]

Accordingly, the Commission describes the situations in which the precautionary principle should be applied:

> In those specific circumstances where scientific evidence is insufficient, inconclusive or uncertain and there are indications through preliminary objective scientific evaluation that there are reasonable grounds for concern that the potentially dangerous effects on the environmental, human, animal or plant health may be inconsistent with the high level of protection chosen for the Community.[18]

Important hereby is that the Commission requires the presence of "reasonable grounds" for considering "potentially dangerous effects." However, crucial terms, such as "scientific uncertainty" are left undefined.[19]

The Communication draws an important distinction between, on the one hand, the decision to make use of the precautionary principle, i.e., the factors that trigger the application of the precautionary principle, and, on the other hand, the decision as to which kind of precautionary measures are to be adopted in each case and under which conditions.[20] Both decisions are eminently political by nature, depending on the level of risk that society is willing to accept, but they must, however, be based on scientific evidence.

> With respect to the factors that may trigger the principle, the Communication states that such triggering "presupposes that potentially dangerous effects deriving from a phenomenon, product or process have been identified, and that scientific evaluation does not allow the risk to be determined with sufficient certainty."[21]

This sentence echoes the definition of the principle established by the European Court of Justice (or seen as such by the legal doctrine) in the 1996 *BSE* case, which referred to the circumstances under which the principle could be triggered (see below).[22] The Commission insists that, in any event, the principle cannot be invoked in order to justify the adoption of arbitrary decisions, and the decision should be based on the strongest possible scientific evaluation.[23]

As such, the Communication provides three prerequisites for invoking the precautionary principle: the identification of possible negative effects, the performance of a scientific evaluation and the existence of scientific uncertainty.[24]

---

[17] Vos, E., De Smedt, K. et al., *Taking stock as a basis for the effect of the precautionary principle since 2000*, 2020. https://recipes-project.eu/.

[18] COM(2000)1, p. 2.

[19] Janssen, A., and Rosenstock, N., 'Handling uncertain risks: An inconsistent application of standards? The precautionary principle in court revisited', *European Journal of Risk Regulation*, 7 (1), 2016, p. 145.

[20] COM(2000)1, p. 12.

[21] COM(2000)1, p. 3.

[22] Case C180/96, UK vs. Commission, para 99.

[23] COM(2000)1, p. 13

[24] COM(2000)1, p. 13.

Hence, some form of scientific evaluation or analysis is mandatory; the fact that in cases of scientific uncertainty no full risk assessment can be carried out[25] (and hence no scientific consensus can be established) does not preclude invocation of the precautionary principle. Hypothetical concerns are not sufficient to trigger the precautionary principle. Grounds for concern that can trigger the precautionary principle are limited to those that are plausible or scientifically tenable. These concerns are based on empirical input and/or modelling outputs that lead to the plausible scientific hypothesis that serious harm appears possible. In the *Pfizer* case[26] (see below), for example, the European General Court required "as thorough a scientific risk assessment as possible, account being taken of the particular circumstances of the case at issue."[27]

Most notably, the Communication states that risk assessment consists of four components, namely hazard identification, hazard characterization, appraisal of exposure and risk characterization, and that an attempt to complete those four steps should be performed before any decision to act is adopted.[28] The risk assessment can give policy makers a more concrete idea of the extent of uncertainty and the means by which it might eventually be solved. In that regard, the Communication prescribes that due attention should also be paid to advice given by a minority fraction of the scientific community, provided that the credibility and reputation of this fraction are recognized.[29]

In a second step, the Communication elaborates on the types of measures to be adopted once the decision to invoke the precautionary principle is taken. In that regard, it specifies from the outset that applying the precautionary principle does not necessarily lead to measures to be designed to produce legal effects that are open to judicial review. A broad range of measures are conceivable, such as funding research programs, informing the public about the potential risk surrounding a certain product or substance, or even in some cases a decision not to take action at all.[30] Furthermore, the Commission establishes guidelines in relation to those precautionary measures, to be followed by policy makers, comprising six components:[31]

---

[25] This refers to the problem that in cases of scientific uncertainty there are at least limits in the availability of data concerning toxicology or exposition of humans. Therefore, a full risk assessment cannot be carried out.

[26] Case T-13/99 Pfizer, para 162.

[27] Vos, E., De Smedt, K. et al., *Taking stock as a basis for the effect of the precautionary principle since 2000*, 2020, p. 83. https://recipes-project.eu/.

[28] COM(2000)1, p. 13 and Annex III. As a reminder and to draw a brief comparison, Article 5(7) SPS Agreement, the clearest reflection of the precautionary principle within the WTO framework, allows for the adoption of precautionary measures only where scientific uncertainty resulting from a lack of available data precludes the performance of a risk assessment. Where drafting an assessment appears to be possible, it is not Article 5(7), but Article 2(2) and Article 5(1) and (2) of the Agreement that apply.

[29] COM(2000)1, p. 16.

[30] COM(2000)1, p. 15.

[31] COM(2000)1, p. 3.

First, the Communication provides that precautionary measures should be proportional to the chosen level of protection.[32] This does not mandate achieving the unrealistic goal of a 'zero-risk' situation; nevertheless, under certain circumstances, the level of uncertainty is such that drastic measures such as bans may be imposed.

Second, the measures should be non-discriminatory in their application.[33]

Third, especially where it proves impossible to adequately characterize the risk due to factors such as lack of data, the measures should be consistent in scope and nature with similar ones already taken in equivalent areas in which all scientific data are available.[34]

Fourth, where appropriate and possible, the Communication states that a cost–benefit analysis should precede the adoption of proposed measures, which implies weighing both economic and non-economic concerns when considering their consequences.[35] In that regard, the Communication specifies that, in line with the Court's case law, the protection of health must take precedence over economic considerations. Once again, having added that any such analysis should only take place "where appropriate and possible" implies that this is a decision for policy makers.[36]

Fifth, the measures should be subject to review in the light of new scientific data. This implies, on the one hand, that even though resulting measures are only intended to be provisional, they should not be revoked until the underlying uncertainties are resolved; and, on the other hand, that scientific research ought to be continued, and that any measures should be subsequently reviewed and potentially modified in light of new scientific findings.[37]

As a final guideline on the adoption of precautionary measures, the Communication prescribes that the latter should be capable of assigning responsibility for producing the scientific evidence necessary for a more comprehensive risk assessment.[38] This is the question of to whom the burden of proving the safety of a product, substance or process should be assigned. In that regard, the Commission indicates that in cases where approval mechanisms were established prior to the development of a product, the burden of proof was placed, a priori, on the manufacturer. Prior approval schemes are common and uncontroversial precautionary measures among EU Member States and also third countries. Despite involving burdensome procedures, they give producers an opportunity—before bringing a product or substance to market—to reconsider whether to proceed with it. If a producer goes through the entire process and its product is ultimately recognized as being safe and commercialized,

---

[32] COM(2000)1, p. 17.
[33] COM(2000)1, p. 18.
[34] COM(2000)1, p. 18.
[35] COM(2000)1, pp. 18 and 19.
[36] COM(2000)1, p. 19.
[37] COM(2000)1, p. 20.
[38] COM(2000)1, p. 21.

it benefits from a situation of sensible legal certainty.[39] Where no prior approval system was established, the Communication prescribes that *ad hoc* precautionary measures could nevertheless be adopted with the effect of reversing the burden of proof onto the producer. According to the Commission, this should not, however, constitute a general rule.[40]

Although the Communication was generally welcomed by the European Council, the Council, the European Parliament, Member States and stakeholders, academic literature published in the early 2000s has been quite critical of the Communication.[41] The main criticisms of the Communication raised in the academic literature are:[42]

- The Communication does not provide a definition of the precautionary principle. Hence, it does not give proper guidance on how the precautionary principle can then best be used;[43]
- Contrary to its declared goals, the Communication does not place meaningful and effective constraints on the application of the precautionary principle. While imposing a 'balancing' activity in deciding whether or not to have recourse to the principle, the communication apparently tipped in favor of adopting preventive measures. Hence, it failed to set a risk threshold triggering its invocation;[44]
- It is naive to assume that decisions based on the precautionary principle can be reversed when new scientific findings become available, as this ignores the problem of technical stigma;[45]
- The Commission does not provide a means to assess and determine which hazards should be prioritized over others in considering the precautionary principle;[46]
- The Communication does not address the problematic issue of risk–risk trade-offs;[47]
- Although in principle the Commission favors cost–benefit analysis, it also argues that this should consider not only the costs to the EU but also those associated

---

[39] Zander, J., *The Application of the Precautionary Principle in Practice: Comparative Dimensions*, Cambridge University Press, New York, 2010, p. 97.

[40] COM(2000)1, p. 4.

[41] Löfstedt, R., 'The precautionary principle in the EU: Why a formal review is long overdue', *Risk Management*, 16(3), 2014, pp. 143–145.

[42] Ibid.

[43] Graham, J., and Hsia, S. 'Europe's precautionary principle: Promise and pitfalls', *Journal of Risk Research*, 5(4), 2002; Majone, G., 'The precautionary principle and its policy implications', *Journal of Common Market Studies*, 40(1), 2002, pp. 89–109; Zander, J., *The Application of the Precautionary Principle in Practice: Comparative Dimensions*, Cambridge, 2010.

[44] McNelis, N., 'EU Communications on the precautionary principle', *Journal of International Economic Law*, 3, 2000, pp. 545–551.

[45] See Graham, J., and Hsia, S., 'Europe's precautionary principle: promise and pitfalls', *Journal of Risk Research*, 5(4), 2002.

[46] Ibid.

[47] Zander, J., *The Application of the Precautionary Principle in Practice: Comparative Dimensions*, Cambridge, 2010.

with a number of non-economic considerations such as public acceptability, leaving the Commission plenty of vague language for interpretation.[48]

Importantly, the literature has criticized the Commission's view that the precautionary principle "pertains to risk management and not also to risk assessment." In this context, we agree that the precautionary principle should be seen as a general governance principle employed throughout the overall process of framing, assessment, evaluation, and management.[49]

In the next section, we examine how the EU institutions have applied the precautionary principle in practice, and whether the Communication has been followed.

Application of the Precautionary Principle Since 2000 by the EU Institutions in Legal Acts

As a baseline for the RECIPES project, an extensive review was conducted of whether and how the precautionary principle has been referred to in the adoption of legal acts since 2000.[50]

In order to understand how the precautionary principle is used in practice by the EU institutions in legal acts, the context of its use must first be understood. Therefore, we first examined how many legal acts employ or refer to the precautionary principle. To this end, we conducted an advanced search on the Eur-Lex portal for the term [precautionary principle] in EU legal acts.

The search returned a total of 135 legal acts, of which 94 (40 regulations, 27 directives and 27 decisions) remained in force in July 2019 which then formed the basis of our further analysis.

A few observations need to be made concerning the number of acts. Firstly, it is clear that the total number of acts found (i.e., 135 acts in a period of 19 years, with 94 of these remaining in force) is quite small in view of the fact that, annually, the EU legislator currently adopts approximately 150 legislative acts and the Commission almost 2000 executive acts (both delegated and implementing). Here, it is important to underline that we did not look into acts that may apply

---

[48] Majone, G., 'The precautionary principle and its policy implications', *Journal of Common Market Studies*, 40(1), 2002, pp. 89–109.

[49] See, e.g., Renn, O., and Dreyer, M. (eds.) *Food Safety Governance*. Springer 2009.

[50] An advanced search on the Eur-Lex portal was performed for the term [precautionary principle] in EU legal acts. The analysis was twofold. We differentiated between instruments used: Regulations, Directives and Decisions; and also between the various types of legal acts: legislative acts adopted by the Council and the European Parliament according to the ordinary legislative procedure, and non-legislative acts adopted by the Commission (since Lisbon-delegated acts and implementing acts). To provide for a complete overview, the period between January 2000 and July 2019 was covered, and the data provided by Eur-Lex, such as directory codes, were exported. The Eur-Lex search allowed for an inductive analysis to understand when and how the precautionary principle is used.

precautionary approaches without explicitly mentioning the precautionary principle. This would be particularly relevant for acts concerning food safety (because of the General Food Law) and the environment (in view of Article 191 TFEU). Consequently, in practice, there may be more situations in which the precautionary principle is being applied. It is therefore acknowledged that the bird's-eye perspective, and hence the search for the term [*precautionary principle*] in legal acts, is an important starting point but that further research is needed to precisely grasp the actual application of the precautionary principle in EU legal acts.

Nevertheless, on the basis of these data it can be observed that, although the precautionary principle is applied to a broad range of topics, it remains primarily a feature of the traditional sectors such as environmental, consumer and health protection. This coincides with the Commission's Communication.[51] Academic research, moreover, shows that for the invocation of the precautionary principle it matters which *Directorate General (DG)* is responsible for addressing the risk issue in question. For example, DG Environment has been found to be more willing to propose precautionary policies than DG Industry.[52]

Importantly, the research also showed that, where the precautionary principle is used as a guiding principle, the reasons for doing so are often poorly explained. Moreover, only six of the legal acts provide a definition of the precautionary principle. The first notable attempt to define the precautionary principle was made in the General Food Law in 2002. Article 7 of this Regulation emphasizes the use of the precautionary principle in response to scientific uncertainty and as part of "risk management." It also clearly establishes the provisional nature of precautionary measures, by stating that they are adopted "pending further scientific information for a more comprehensive risk assessment" (Art. 7(1)) and are subject to review "within a reasonable period of time" (Art. 7(2)). The trigger for the use of the principle, specified here as "possibly harmful effects on health," must necessarily remain imprecise. This corresponds to the definition and criteria established in the Commission's 2000 Communication.

Article 7, General Food Law (Regulation 178/2002)

1. In specific circumstances where, following an assessment of available information, the possibility of harmful effects on health is identified but scientific uncertainty persists, provisional risk management measures necessary to ensure the high level of health protection chosen in the Community may be adopted, pending further scientific information for a more comprehensive risk assessment.
2. Measures adopted on the basis of paragraph 1 shall be proportionate and no more restrictive of trade than is required to achieve the high level of health protection chosen in the Community, regard being had to technical and economic feasibility and other factors regarded as legitimate in the matter under consideration. The measures shall be reviewed within a reasonable period of time,

---

[51] COM (2000) 1, p. 8.

[52] Tosun, J., and Pesendorfer, D., 'EU environmental policy under pressure: Chemicals policy change between antagonistic goals?', *Environmental Politics*, 15(1), 2006, p. 101.

depending on the nature of the risk to life or health identified and the type of scientific information needed to clarify the scientific uncertainty and to conduct a more comprehensive risk assessment.

As this is one of the rare instances in which a clear definition of the precautionary principle is spelled out, it is not surprising that references to the General Food Law are also contained in other legal acts. This is the case in two other food-related Regulations as well as the Regulation on plant protection products.[53]

Besides this, the precautionary principle is identified only in one other act of general application revealed by our search: Article 2 of a Council Decision on the Protocol on the Implementation of the Alpine Convention in the field of transport (Transport Protocol). Compared to the definition laid down in the General Food Law, the threshold for triggering the precautionary principle seems slightly higher in the Transport Protocol, which makes reference to "serious irreversible effects on the health and the environment," albeit indicating that this also means "potential harmfulness." The General Food Law, in contrast, departs from "potential harmful effects on health."

With regard to the requirement of scientific uncertainty, both legal acts show similarities. The Transport Protocol refers to situations where "research has not yet strictly proven the existence of a cause-and-effect relationship" between substances and potential harm, whereas the General Food Law mentions that "scientific uncertainty persists." The measures to be taken in such situations are those "intended to avoid, control or reduce effects" of such harm (Transport Protocol) or "risk management measures" necessary to ensure health protection (General Food Law).

The formulation of the action to be taken differs slightly. Whilst the Transport Protocol links with the triple negative formulation of the Rio Declaration in stating that measures "should not be postponed" by reference to uncertainty, the General Food Law holds that measures "may be adopted."

Moreover, the General Food Law clearly states that the measures are of a "provisional" nature and "pending further scientific information"; in contrast, no such indication is given in the Transport Protocol definition, although this is one of the requirements foreseen in the 2000 Communication.

It can therefore be said that there is no single definition of the precautionary principle in EU legal acts. The EU General Food Law regulation is quite exceptional in that it expressly defines the precautionary principle for application in that sector. EU environmental legislation provides no equivalent definition, although the TFEU (Treaty on the Functioning of the European Union) directly refers to the precautionary principle as a basis for EU environmental policy. This has left the precautionary principle open to interpretation within the environmental policy area.

On the one hand, this is advantageous as it allows for flexibility and the possibility to adapt to individual needs of environmental problems. Commentators have

---

[53] Reg 2015/2283, Art. 12+18, Reg 1107/2009, Art. 13 (plant protection products), Reg 609/2013, Art. 5 (food intended for infants and young children, food for special medical purposes, and total diet replacement for weight control).

generally viewed the lack of general definition of the precautionary principle at EU level positively, as the principle's application differs across the range of policies and must be context specific. Quite evidently, this has led to different approaches and interpretations. This is why both the literature and the Commission, instead of giving a firm definition, prefer to speak of the "constituent parts"[54] of the precautionary principle.

Whilst it is neither necessary nor possible to strive for a general legal definition of the precautionary principle in EU law, it is of crucial importance that procedures for the application of the principle, such as the ways in which risk assessments are performed, the transparency in dealing with uncertainties, and how different strengths of evidence for action are evaluated and chosen, are similar and predictable.[55]

Our analysis of legal acts reveals, however, that the invocation of the precautionary principle is diverse and seems to leave gaps with regard to a precise definition and application of the principle. Scholarly analyses confirm these findings. Garnett and Parsons reviewed a small sample of Directives and Regulations and concluded that the precautionary principle was applied differently in EU law, and with very little consistency across cases regarding the conditions for taking precautionary action and the basis for imposing regulation.[56] Their review of a limited numbers of legal acts (4 Regulations, 4 Directives and 3 Decisions of the EU legislators and the Commission) reveals that application of the precautionary principle varies in strength from weak to moderate and strong precaution.

The above shows that the guidelines laid down in the Commission's Communication are not followed consistently in legal practice. These findings could cast doubt on the impact of the 2000 Communication. Löfstedt, for example, argues that the Communication has been little used in practice, and calls for a review of the Communication.[57] These findings, moreover, largely correspond to those of a study on the use of the precautionary principle in EU Environmental policies performed by Milieu for the European Commission in November 2017. The latter study, for example, shows that certain aspects, such as methodologies for assessing risk and the question of when precautionary action is required, vary across the different environmental sectors. This can be explained by the differing, content-specific approaches taken.[58] However, as indicated above, a coherent application of procedures for the application of the principle, such as the ways in which risk assessments

---

[54] COM(2000)1, p. 12.

[55] European Commission, Study on the precautionary principle in EU environmental policies, Final Report – Study, Milieu, November 2017, pp. 93–94.

[56] Garnett, K., and Parsons, D. J., 'Multi-Case Review of The Application of The Precautionary Principle in European Union Law and Case Law.' *Risk Analysis: An official publication of the Society for Risk Analysis*, 2017, p. 37.

[57] Löfstedt, R., 'The precautionary principle in the EU: Why a formal review is long overdue', *Risk Management*, 16(3), 2014, p. 149.

[58] European Commission, Study on the precautionary principle in EU Environmental Policies, Final Report – Study, Milieu, November 2017, DOI 10.2779/58953 KH-07-17.198-EN-N.

are performed, transparency in dealing with uncertainties, and how different strengths of evidence for action are evaluated and chosen, are of fundamental importance to avoid tensions between the precautionary principle and innovation.

We can also refer here to the Ombudsman's view of the precautionary principle as a principle of good administration.[59] This consideration may link up, on one side, with the acceptance of a lack of a general legal definition of the precautionary principle in EU law and, on the other side, with the view that one should look more closely into establishing similar and predictable procedures for its application.

Application of the Precautionary Principle by the European Courts in Court Rulings from 2000 to 2019

Besides examining how the precautionary principle has been referred to in the adoption of legal acts since 2000, the RECIPES project also examined how the European Court of Justice has dealt with the precautionary principle. In this regard, we performed a literature review, a review of seminal Court cases and a quantitative analysis.

As a preliminary matter, it should be noted that the Court of Justice of the European Union (CJEU) interprets EU law to ensure it is applied consistently across all EU Member States, and settles legal disputes between national governments and EU institutions. It can also be used by individuals, companies, or organizations to take action against an EU institution if they feel it has infringed their rights. The CJEU is divided into **two courts:** the Court of Justice, which deals with requests for preliminary rulings from national courts, actions for annulment and appeals; and the General Court, which rules on actions for annulment brought by individuals, companies and, in some cases, governments of EU Member States.[60]

Definitions of the Precautionary Principle

As the 2000 Communication did not provide a definition of the precautionary principle, CJEU case law is crucial in determining when, how and by whom the precautionary principle may be relied upon in the EU legal order.[61] However, the courts visibly struggle with this role and certain inconsistencies have arisen.

---

[59] See in this regard Para 10 of Decision in case 12/2013/MDC on the practices of the European Commission regarding the authorization and placing on the market of plant protection products (pesticides), https://www.ombudsman.europa.eu/en/decision/en/64069; and the Decision in case 23/2018/SRS on how the European Commission updates EU rules on chemical testing when alternative test methods are identified; https://www.ombudsman.europa.eu/en/decision/en/109429.

[60] See, e.g., Türk, A., *Judicial Review in EU Law*, Edward Elgar, 2010.

[61] See Alemanno, A., 'The shaping of the precautionary principle by European courts: From scientific uncertainty to legal certainty', *Bocconi Legal Studies Research Paper*, 2007.

References to the precautionary principle in case law of the Court of Justice and the General Court between 2000 and 2019 are generally considerably more detailed than references in legal acts. In total, the Eur-Lex search for the expression [*precautionary principle*] yielded 147 results. This includes judgments by both the General Court and the Court of Justice in procedures under articles 260, 263, 267 and 340 TFEU. The codes used by Eur-Lex indicate that the subject areas covered in these judgments are similar to the findings in legislation. The codes for environment (70 instances), approximation of laws (53), agriculture and fisheries (41) as well as health (21) were used most often.[62]

The courts have made important contributions to the understanding of the precautionary principle in milestone cases, namely: *Alpharma* (2002), *Artegodan* (2002), *Pfizer* (2002), *Solvay Pharmaceuticals* (2002), *Paraquat* (2007), *Gowan* (2009), *SPCM* (2009), *Afton* (2010), *Bayer CropScience* (2018) and *Confederation Paysanne* (2018). We refer to these cases to show tendencies in the EU courts' application and interpretation of the precautionary principle.

The courts have given various definitions of the precautionary principle, which have been formalized over time. Generally, we can observe that the courts use three different formulations of the precautionary principle, one of which is further differentiated depending on whether the principle is invoked by the Commission or a Member State.

Above, we already reported on the Court's definition of the precautionary principle to be applied by the EU institutions in the *BSE* case, which has been repeated as a standard formulation in many other cases:

> Where there is uncertainty as to the existence or extent of risks to human health, protective measures may be taken without having to wait until the reality and seriousness of those risks become fully apparent.[63]

A similar formulation is used for the Member States:

> It is clear that such an assessment of the risk could reveal that scientific uncertainty persists as regards the existence or extent of real risks to human health. In such circumstances, it must be accepted that a Member State may, in accordance with the precautionary principle, take protective measures without having to wait until the reality and seriousness of those risks are fully demonstrated.[64]

In other cases, the courts stated that:

> The precautionary principle allows the adoption of provisional risk management measures necessary to ensure a high level of health protection when, following an assessment of

---

[62] As in the case of legislation, several descriptors were sometimes used for one document. Consequently, the number of descriptors does not correspond to the total number of judgments.

[63] Case C-180/96, para 99. See later cases C-343/09, para 62; C-77/09, para 73; T-429/13, para 110; T-13/99, para 139; T-70/99, para 152; T-141/00, para 185; C-269/13, para 57; T-108/17, para 281; T-584/13, para 59; T-257/07, para 68; C-78/16, para 47; T-392/02, para 122; T-817/14, para 51; T-31/07, para 135; C-151/17, para 38; C-157/14, para 81; T-334/07, para 116; C-477/14, para 47; C-236/01, para 111.

[64] See cases C-41/02, para 52; C-282/15, para 60; C-446/08, para 67; C-333/08, para 91.

available information, the possibility of harmful effects on health is identified but scientific uncertainty persists [pending further scientific information].[65]

And

> Where it proves to be impossible to determine with certainty the existence or extent of the alleged risk because of the insufficiency, inconclusiveness or imprecision of the results of studies conducted, but the likelihood of real harm to public health persists should the risk materialise, the precautionary principle justifies the adoption of restrictive measures.[66]

These definitions consistently point to scientific uncertainty as the main factor for invoking the precautionary principle, allowing for restrictive measures. Moreover, case law that was issued after the 2000 Communication makes mention of the provisional character of the risk management measures whilst also pointing out that precautionary action may only be taken following an initial assessment of the available information. Our analysis reveals that no explicit evidence of risks is necessary in order for the regulator to rely on the precautionary principle.

Limited Judicial Review of the Precautionary Principle

We reviewed cases in which the precautionary principle was invoked, in order to define commonalities with respect to the Courts' review of these cases. The analysis of the cases selected shows that the courts' review is limited to a small number of potential factors. These factors can be broadly divided into three categories: *first*, the reasons for triggering the use of the principle; *second*, the considerations that the regulator must take into account in the decision-making phase; and *third*, the requirements that any resulting measures must comply with.

We will now discuss each of these categories.

Review of Reasons for Triggering the Use of the Precautionary Principle

First, the Courts must decide whether the invocation of the precautionary principle was justified. Here, the Court attempts to define the precautionary principle and examines the elements that justify the use of the precautionary principle. In this stage, the Court only reviews whether the conditions for applying the precautionary principle are fulfilled (i.e., the sufficiency of scientific uncertainty) and ensures that the regulator does not base its decision on purely hypothetical risks.

---

[65] See cases T-257/07, para 67; C-282/15, para 54; C-111/16, para 44; C-192/01, para 49.

[66] See cases C-343/09, para 61; C-77/09, para 76; T-429/13, para 119; C-192/01, para 52; C-95/01, para 48; C-41/02, para 54; C-333/08, para 93; C-446/08, para 70; T-31/07, para 142; C-269/13P, para 58; C-157/14, para 82; T-817/14, para 51; C-477/14, para 47; C-78/16, para 47; C-78/16, para 47; C-282/15, para 57; T-584/13, para 68; C-151/17, para 38; C-489/17, para 58; T-108/17, para 282.

In the milestone case involving *Pfizer*, the Court of First Instance (now General Court) for the first time discussed the interpretation and the correct application of the precautionary principle and defined the *conditions for triggering* the application of the precautionary principle.[67]

According to the judgment:

> in case of scientific uncertainty as to the existence of a risk to human health, the EC institutions as well as the Member States may invoke the precautionary principle in order to adopt protective measures, in spite of the fact that a proper risk assessment showing conclusive scientific evidence cannot be conducted.[68]

More specifically, the *factors for triggering* the precautionary principle are further described with regard to two different elements. First, the requirement that the risk on which the action is based may not be hypothetical is emphasized. As has been demonstrated previously, this is one of the elements the Court is indeed willing to review.

Second, the Court moreover applies the definition of the precautionary principle to very specific authorization procedures and emphasizes above all the need for *'solid evidence'*. Thus, the Court requires sufficient evidence in order to conclude that there is insufficient scientific information about the prevalence of certain risks. This is what van Asselt and Vos have referred to as the 'uncertainty paradox'.[69] Whilst an insufficient amount of evidence can, in principle, be a reason for the Court to review decisions, this had an effect in the decisions analyzed only where procedural errors were made. The exact level of uncertainty needed is difficult to assess and therefore, in practice, only subject to a very limited review.

In procedural terms, it is important for the regulator to conduct a risk assessment in order to provide the required level of 'solid evidence'. In practice, however, this requirement is not always fulfilled. In several cases (*Alpharma, Solvay*) no risk assessment was performed, and the Court did not reprimand the Commission or the Council for not conducting a risk assessment. Instead, the Court "acted as a super risk assessor," whereas it ought to have determined whether the risk manager conducted a risk assessment and whether this was done according to the procedural requirements. In its place, the Court *constructed uncertainty as the absence of full safety*.[70]

---

[67] Janssen, A., and van Asselt, M., 'The precautionary principle in court – An analysis of post-Pfizer case law', in van Asselt, M., Versluis, E., Vos, E. (eds.), *Balancing between Trade and Risk: Integrating Legal and Social Science Perspectives*, London, UK: Routledge, 2013, p. 199.

[68] Case T-13/99, Pfizer Animal Health SA v. Council, 'Pfizer', 2002.

[69] van Asselt, M., and Vos, E., 'The precautionary principle and the uncertainty paradox', *Journal of Risk Research*, 9(4), pp. 313–336.

[70] Janssen, A., and van Asselt, M., 'The precautionary principle in court – An analysis of post-Pfizer case law', in van Asselt, M., Versluis, E., Vos, E., (eds.), *Balancing between Trade and Risk: Integrating Legal and Social Science Perspectives*, London, 2013, p. 213.

The academic literature emphasizes that other cases show the same lack of a proper risk assessment.[71] In *Afton*, the Commission did not conduct a risk assessment to determine the negative impact of MMT (methylcyclopentadienyl manganese tricarbonyl, a fuel additive) on pollution abatement techniques.[72] In *Bayer CropScience*, the Court accepted expert consultations as a sufficient form of risk assessment.[73] Furthermore, both Alemanno and Zander argue that in both the *Paraquat* and *Gowan* court cases the European court has supported the use of the precautionary principle to ban substances without robust scientific evidence of their potential effects (and in *Paraquat* the Court even found that the Commission *must* act in a precautionary manner).[74]

However, in various cases the Court states that:

> In the domain of [human health], the existence of solid evidence which, while not resolving scientific uncertainty may reasonably raise doubts as to the safety of a substance justifies, in principle, [the refusal to include that substance…]. The precautionary principle is designed to prevent potential risks.[75]

The Court has therefore repeatedly held that:

> The risk assessment cannot be based on purely hypothetical considerations.[76]

Van Asselt and Vos argue that, in *Pfizer*, the Court equated scientific uncertainty with diverging opinions and thereby constructed its own definition of uncertainty.[77] They highlight the possibility that, in this manner, the precautionary principle might be applied whenever one qualified scientist holds a diverging opinion.[78]

Janssen and van Asselt (2013)[79] examined post-Pfizer case law to determine whether the problematic ruling of the Court in *Pfizer* had set a precedent. They identified several tensions and inconsistencies in the Court's rulings on *Pfizer*, *Alpharma*, *Artegodan* and *Solvay Pharmaceutical* (all during 2002), both with

---

[71] Janssen, A., and Rosenstock, N., 'Handling uncertain risks: An inconsistent application of standards? The precautionary principle in court revisited', *European Journal of Risk Regulation*, 7 (1), 2016, p. 150.

[72] Ibid.

[73] Ibid, p. 150. See also Alemanno, A., 'The science, law and policy of neonicotinoids and bees: A new test case for the precautionary principle', *European Journal of Risk Regulation*, 4, 2013.

[74] Zander, J., *The Application of the Precautionary Principle in Practice: Comparative Dimensions*, Cambridge, 2010, p. 130. See also Alemanno, A., Annotation of European Court of Justice case C-79/09, Gowan Comércio Internacional e Serviços Lda v. Ministero Della Salute (Precautionary Principle), *Common Market Law Review*, 48, 2011 pp. 1329–1348.

[75] Established in T-141/00, para 192. See further cases C-236/01, para 113; T-392/02, para 129; T-326/07, para 166; T-334/07, para 180; T-71/10, para 75, T-429/13, para 116.

[76] See cases C-192/01, para 49; C-41/02, para 52; C-269/13P, para 58; T-584/13, para 65.

[77] van Asselt, M., and Vos, E., 'The precautionary principle and the uncertainty paradox', *Journal of Risk Research*, 9(4), 2006.

[78] Ibid.

[79] Janssen, A., and van Asselt, M., 'The precautionary principle in court – An analysis of post-Pfizer case law', in van Asselt, M., Versluis, E., Vos, E. (eds.), *Balancing between Trade and Risk: Integrating Legal and Social Science Perspectives*, London, UK: Routledge, 2013, p. 199.

respect to the prerequisites for invoking precautionary principle and also the measures eventually taken. Furthermore, Janssen and Rosenstock criticize the Court's lack of vision on how to deal with uncertainty and precaution.[80] Janssen and van Asselt hold that, in *Pfizer*, the General Court used scientific disagreement as a way of constructing uncertainty about the risk in question. The Court referred to diverging opinions between experts, which was subsequently used to legitimize the application of the precautionary principle.[81]

Moreover, in *Alpharma*, uncertainty was not only constructed through a lack of scientific consensus; moreover, the Court also argued in terms of analogy with other antibiotics. In this case, there were no risk assessments performed on the specific substance bacitracin zinc. The Court, however, ruled that "all antibiotics and all nitrofurans have similar characteristics and should be treated in the same way."[82] As Janssen and van Asselt argued, this argumentation entails that substance-specific characteristics are no longer relevant to risk assessments, and that commonalities suffice.[83] Analogy was also applied in the *Solvay* case involving the antibiotic Nifursol and subsequently in the cases of both *Gowan* and *Bayer CropScience*. Janssen and Rosenstock argue that, with this approach to establishing uncertainty, the precautionary principle could easily become a tool to prohibit marketing of products.[84]

Review of Considerations Taken Into Account in the Decision-Making Phase

Second, the decision-making process itself requires a complicated assessment of scientific data on the one hand and societal preferences on the other, both of which are difficult issues for a court to review. The EU courts indeed declare in many cases their review to be limited to manifest errors.

The courts operate in this decision-making phase under the precautionary principle another set of formulations. Interestingly, the courts state in various cases that the precautionary principle can 'require' the institutions to take action. This is in sharp contrast to other cases in which the courts held that the precautionary principle "may warrant the adoption of a restrictive measures by an institution" but "does

---

[80] Janssen, A., and Rosenstock, N., 'Handling uncertain risks: An inconsistent application of standards? The precautionary principle in court revisited', *European Journal of Risk Regulation*, 7 (1), 2016.

[81] Janssen, A., and van Asselt, M., 'The precautionary principle in court – An analysis of post-Pfizer case law', in van Asselt, M., Versluis, E., Vos, E. (eds.), *Balancing between Trade and Risk: Integrating Legal and Social Science Perspectives*, London, 2013, p. 199.

[82] Ibid, p. 207.

[83] Ibid, p. 207.

[84] Janssen, A., and Rosenstock, N., 'Handling uncertain risks: An inconsistent application of standards? The precautionary principle in court revisited', *European Journal of Risk Regulation*, 7 (1), 2016, p. 146.

not require it to do so."[85] The Commission has also stated that, in its view, the precautionary principle "does not... oblige the Community institutions to follow all scientific opinion without any margin for assessment."[86]

In other cases, the courts' formulations are variants of the same starting point: they state that the regulator must follow certain steps when applying the precautionary principle. The regulator must identify the risk and then conduct an assessment relying on relevant data. In this regard, the General Court views that risk management measures can be taken on the basis that:

> Within the process leading to the adoption by an institution of appropriate measures to prevent specific, potential risks to public health, safety and the environment by reason of the precautionary principle, three successive stages can be identified: firstly, identification of the potentially adverse effects arising from a phenomenon; secondly, assessment of the risks to public health, safety and the environment which are related to that phenomenon; thirdly, when the potential risks identified exceed the threshold of what is acceptable for society, risk management by the adoption of appropriate protective measures.[87]

This is also confirmed by both courts, where they state that:

> The correct application of the precautionary principle presupposes, first, identification of the potentially negative consequences for health of the substances or foods concerned, and, second, a comprehensive assessment of the risk to health based on the most reliable scientific data available and the most recent results of international research.[88]

These step-by-step conditions offer the courts another, much clearer, possibility for review. Whilst substantive decisions are difficult to review, these seemingly clear procedural issues can be assessed in a comprehensive way and have, as seen in the *BASF* case, led the General Court to annul a Commission decision.[89]

Importantly, the General Court views that measures to protect human health and safety and the environment take precedence over economic interests.

> The precautionary principle is a general principle of EU law requiring the authorities in question, in the particular context of the exercise of the powers conferred on them by the relevant rules, to take appropriate measures to prevent specific potential risks to public health, safety and the environment by giving precedence to the requirements related to the protection of those interests over economic interests.[90]

---

[85] Case T-108/17 ClientEarth v European Commission [2019], ECLI:EU:T:2019:215, para 284.

[86] Case T-304/01 Julia Abad Pérez and Others v Council of the European Union and Commission of the European Communities [2006], ECLI:EU:T:2006:389, para 80.

[87] See cases T-429/13, para 111; T-257/07, para 69; T-31/07, para 136; T-584/13, para 60.

[88] See cases C-343/09, para 60; C-77/09, para 75; C-192/01, para 51; C-41/02, para 53; C-333/08, para 92; C-446/08, para 69; C-282/15, para 56; C-489/17, para 57; T-108/17, para 281.

[89] Case T-584/13 BASF Agro BV and Others v European Commission [2018], ECLI:EU:T:2018:279.

[90] Established in T-141/00, para 184. See further cases T-429/13, para 109; T-141/00, para 184; T-392/02, para 121; T-584/13, para 58; T-817/14, para 51; T-257/07, para 66; T-433/13, para 102; T-31/07, para 134.

Review of the Requirements that Measures Resulting
from the Precautionary Principle Must Comply With

The measures resulting from the decision-making process are mainly reviewed by the courts to ensure that they are proportional and do not aim, unrealistically, for a "zero-risk approach." The courts impose clear conditions: non-discrimination, proportionality and objectivity are central in the judicial review of decisions: "Such measures must not be allowed unless they are non-discriminatory, proportional and objective."[91]

Moreover, the courts repeatedly emphasized that the regulator should not aim for a "zero-risk approach." Studies have confirmed that *proportionality* is subject to a more thorough review than other criteria, such as the need for new scientific data.[92] Rogers highlights that this should not be surprising, considering that the topic of proportionality has been long discussed in European circles, while the fact that precautionary actions should be subject to review has, to date, never been tested in the courts.[93] The proportionality principle is well established in EU law, and the courts have considerable practice in applying it,[94] so much so that, in *Pfizer*, the fourth criterion (action should be subject to costs and benefits of the proposed action) was effectively subsumed by the courts under the proportionality test.[95]

Furthermore, whilst the courts emphasize that a *zero-risk policy* is not acceptable, it has been repeatedly criticized on this issue. Another requirement with regard to the outcome, which is clearly mentioned in the Communication but significantly less subject to judicial review, is the requirement to *review the measure* in light of new scientific data. It appears that in various cases the courts have ignored the temporary nature of precautionary measures. However, such re-evaluations are necessary to prevent precautionary measures becoming permanent contrary to the indications of new scientific evidence. Instead of demanding a substantive review of the latest scientific findings, the General Court found in *Solvay*[96] that an administrative review is sufficient when deciding on precautionary measures. Consequently, "by not insisting on a new risk assessment of the substances, the Court disregards the temporary character of the precautionary principle."[97]

---

[91] See cases C-77/09, para 76; T-429/13, para 117; C-192/01, para 53; T-392/02, para 125; C-333/08, para 93; C-446/08, para 67; T-71/10, para 76; T-817/14, para 51; C-477/14, para 48; C-78/16, para 48; C-282/15, para 57; T-584/13, para 68; C-489/17, para 58; T-108/17, para 282.

[92] Rogers, M., 'Risk management and the record of the precautionary principle in EU case law', *Journal of Risk Research*, 14 (4), 2011, pp. 467–484.

[93] Ibid, p. 478.

[94] See, e.g., Tridimas, T., *The General Principles of EU Law*, Cambridge University Press, 2006.

[95] Rogers, M., 'Risk management and the record of the precautionary principle in EU case law', *Journal of Risk Research*, 14 (4), 2011, p. 480.

[96] Case T-392/02.

[97] See Janssen, A., and van Asselt, M., 'The precautionary principle in court – An analysis of post-Pfizer case law', in van Asselt, M., Versluis, E., Vos, E. (eds.), *Balancing between Trade and Risk: Integrating Legal and Social Science Perspectives*, London, 2013, p. 213.

The General Court's ruling in the *Artegodan* case, however, differed from that in *Solvay*. The General Court explicitly argued that old data that had been used in previous assessments may not constitute a sufficient basis upon which to establish scientific uncertainty in the present.[98]

Conclusions on the European Courts' Application of the Precautionary Principle

In sum, the academic literature reveals several inconsistencies in the courts' rulings in dealing with uncertain risks, which has led to several problematic patterns in which the precautionary principle is effectively utilized as a tool for risk management. The academic literature details a number of recurring issues.

First, with respect to the prerequisites for invoking the precautionary principle, the EU courts in some cases define uncertainty simply as differing scientific opinions, or a lack of consensus between experts. This is a delicate issue that could open up the prospect of protectionism, since differing scientific opinions can be found in many scenarios involving uncertain risks. Therefore, requirements as to the production of such scientific opinions that form the basis of regulatory measures, which experts participate, etc., become of key importance here.

Next, in various cases, the courts have accepted the use of the precautionary principle in the absence of proper scientific evidence. The courts ruled that the possibility of a risk, the absence of zero risk, or the lack of information establishes uncertainty and risk, and is therefore sufficient legal basis for precautionary measures. These are very low thresholds for invoking precautionary measures, since any form of uncertainty requires assessment of risk (which may have positive or negative outcomes). In some cases, the academic literature has criticized the courts for seemingly accepting even inadequate risk assessments.[99] Moreover, to various extents in the different cases: no risk assessments were performed by independent bodies, risk assessments were ignored, and analogy between substances and expert consultations were deemed sufficient.[100]

---

[98] Ibid, pp. 212–213.

[99] See Löfstedt, R., 'The precautionary principle in the EU: Why a formal review is long overdue', *Risk Management*, 16(3), 2014, p. 147. See also Zander, J., *The Application of the Precautionary Principle in Practice*, Cambridge University Press, New York, 2010 and Alemanno, A., 'Annotation of European Court of Justice case C-79/09, Gowan Comércio Internacional e Serviços Lda v. Ministero Della Salute (Precautionary Principle)', *Common Market Law Review*, 48, 2011, pp. 1329–1348.

[100] Janssen, A., and Rosenstock, N., 'Handling uncertain risks: An inconsistent application of standards? The precautionary principle in court revisited', *European Journal of Risk Regulation*, 7 (1), 2016, p. 150.

It has been argued that the review of proportionality is often insufficiently strict.[101] Moreover, it has been asserted that the courts have also disregarded the temporary nature of risk measures by failing to insist on new risk assessments or ignoring new information, despite the requirement that each case must be reviewed based on the latest scientific evidence available. Rogers so holds that EU courts could make a provisional or interim order pending further research but, so far, they have not done so.[102]

Comparing the courts' case law with the Commission 2000 Communication, we can note that the courts do pay attention to the 2000 Communication, but—from analyzing the milestone cases—not consistently. Consequently, it appears that, in particular, the requirements for a robust risk assessment or cost–benefit analysis are not met in all cases.

Furthermore, the sample analyzed in this work is too small to provide definite conclusions, but it seems that in most cases the courts agree with a ban or upholds restrictions. It seems that the courts generally adopt a moderate to strong interpretation of the precautionary principle. These findings support those in the academic literature, in that it seems that the 2000 Communication does not provide sufficient guidance on the application of the precautionary principle. On the other hand, it should be remembered that the 2000 Communication is a non-binding guidance document and that the courts are not bound to apply the criteria proposed by the Commission.

Concluding Remarks

Our contribution highlights that the precautionary principle is employed as a principle of EU law, while there is the lack of a single definition. Turning to the practical application of the precautionary principle in the EU, our legal analysis reveals that, consequently, the criteria for precautionary action, as described in the Commission's Communication on the precautionary principle, are not applied consistently by EU policy makers or the EU courts.

The lack of a single definition of the precautionary principle in EU law can be viewed as advantageous, as it leaves ample room for flexibility and establishing *ad hoc* solutions to context-specific problems. Quite evidently, this has led to different approaches and interpretations. Moreover, it leaves the EU courts with the difficult task of reviewing precautionary measures adopted by the EU institutions. Although these courts have formulated definitions and requirements for applying the precautionary principle, it is also clear that at times they are inconsistent and

---

[101] Alemanno, A., Annotation of European Court of Justice case C-79/09, Gowan Comércio Internacional e Serviços Lda v. Ministero Della Salute (Precautionary Principle), *Common Market Law Review*, 48, 2011, pp. 1329–1348.

[102] Rogers, M., 'Risk management and the record of the precautionary principle in EU case law', *Journal of Risk Research*, 14 (4), 2011, p. 481.

visibly struggle to review measures involving scientific uncertainty. Where, para-doxically, regulators tend to ask for greater scientific certainty in resolving cases of scientific uncertainty, science increasingly appears in the courtroom.[103] Yet, unless there are procedural mistakes or manifest errors of assessment, the EU courts have often been quite reluctant to annul precautionary decisions in view of the large dis-cretion the Commission has in such cases.

To be sure, decisions involving the precautionary principle must often strike a delicate balance between risk assessments on the one hand and societal risk toler-ance on the other. In addition to reasons of balance of powers and the rule of law, it is therefore quite understandable that the courts leave the EU legislator and the Commission much discretion to do so.[104]

Furthermore, the courts seem largely disinterested in reviewing the supposedly temporary nature of precautionary measures. Some judgments seem to overlook the "dynamics of science," e.g., new scientific evidence questioning prior judgement. The requirement set forth in the Commission's Communication—that precaution-ary measures should be provisional, pending a reduction in the scientific uncer-tainty—is still to be seriously addressed by the EU courts.[105]

Ultimately it is important to acknowledge the limits of science in complex judi-cial decision-making and to realize that "any effort to bring better science into the courtroom must respect the courts' constitutionally specified role," "even if doing so means, from time to time, what is, from a scientific perspective, an incorrect result."[106] Hence, judges should remain judges; they should not become "amateur scientists."[107] Courts should therefore leave responsibility for science-based deci-sion making to decision makers, as advised by scientists.

In the EU context, it might be helpful for decision makers (the EU legislator, the Commission, or the Member States) to receive more guidance, perhaps in the form of a revised Commission communication. Such a revision could clarify—without losing track of the need for flexibility to adapt to the specific needs of individual problems—the role of science, and give guidance to regulators on how to deal with scientific uncertainty in decision making. This could be particularly helpful in cases where such certainty cannot presently be provided by current scientific knowledge

[103] Vos, E., 'The European Court of Justice in the face of scientific uncertainty and complexity' in: M. Dawson, B. de Witte and E. Muir, *Judicial Activism at the European Court of Justice*, Edward Elgar, 2013, pp. 142–160.

[104] Vos, E., De Smedt, K. et al., *Taking stock as a basis for the effect of the precautionary principle since 2000*, 2020, p. 91. https://recipes-project.eu/.

[105] See Rogers, M., 'Risk management and the record of the precautionary principle in EU case law', *Journal of Risk Research*, 14 (4), 2011, p. 481.

[106] As American Justice Breyer said in: General Elec. Co. v. Joiner, 118 S. Ct. 512, 520 (1997) (Breyer, J., concurring), quoted in Scallen, E., and Wiethoff, W., The ethos of expert witnesses: Confusing the admissibility, sufficiency and credibility of expert testimony, *Hastings Law Journal*, 49(4), 1998, p. 1167.

[107] Scallen, E., and Wiethoff, W., The ethos of expert witnesses: Confusing the admissibility, suf-ficiency and credibility of expert testimony, *Hastings Law Journal*, 49(4), 1998, p. 1145.

in avoiding a quest for ever greater scientific certainty, thereby easing a perceived tension between precaution and innovation.

This is of particular importance in striking an appropriate balance between, on the one hand, concerns for health, safety, and environmental protection and, on the other hand, economic interests. It is clear that the intricacies of dealing with precautionary measures involving science and scientific uncertainty in courtrooms remain delicate and require further scholarly attention.

Acknowledgement The RECIPES project has received funding from the European Union's Horizon 2020 research and innovation programme under grant agreement No 824665.

Part III
Science and Responsibility—The Scientists' Voices

The four essays in this third part of the book take a stand, each for itself. This can be done in a very personal way, such as Klaus Fuchs-Kittowski's essay, or in an institutional-programmatic way, like VWD's position paper on digitization. The question in both cases is whether the formal apparatus of science so obscures the human that research itself can become inhuman, a question which can play a role in the context of artificial intelligence. The other two essays deal with the current topics of misinformation ('fake news') on the one hand and co-production of knowledge on the other.

Chapter 9: Rainer E. Zimmermann's contribution, "Between *Parrhesia* and *Fake*: Scientific Responsibility Today," is the prelude to this third part concerning scientists' voices. Zimmermann emphasizes that scientific discourse presupposes free speech, but at the same time must always allow itself to be subjected anew to the test of truth.

Chapter 10: Klaus Fuchs-Kittowski's contribution, "The Responsibility of Science for Guaranteeing Human Rights in the Fight Against Human Degradation, Racism and Anti-Semitism," is both emphatic and provocative. Fuchs-Kittowski doubts that truth is of any use unless it is guided by humanitarian values.

Chapter 11: The contribution of the VDW (The Federation of German Scientists, Hartmut Graßl et al.) is a position paper on the use and development of digitization in science and society. The VDW's core demand is a review of the image of humanity that is associated with digitization projects (which may lead to project adjustments).

Chapter 12: In "Heritage Requires Citizens' Knowledge," Heike Oevermann et al. (COST) reflect on how Responsible Research and Innovation (RRI) can be implemented in an EU project. They emphasize that the real involvement of citizens is difficult and prerequisite-rich.

About the authors: *Rainer E. Zimmermann* (born 1951) a scholar whose breadth of knowledgeable is rarely found in today's professionalized science—a trained physicist, teaching as a philosopher, and working as a philologist. He is a life member of Clare Hall and a feisty chairman of scholarly societies. *Klaus Fuchs-Kittowski* (born 1934) is the grandson of Emil Fuchs (1874–1971), a German theologian who opposed the Nazi regime. Emil Fuchs was father of the atomic spy Klaus Fuchs (1911–1988), who emigrated to the UK, studied physics there, and betrayed details of the American nuclear weapons program (Manhattan Project) to the Soviet Union during World War 2 and until 1949. The *VDW* (Federation of German Scientists), was founded in 1959 as a scientific conscience, thereby initially in opposition to the development of nuclear weapons and today in the context of climate change. The current director, Hartmut Graßl (born 1940), was director of the German Max Planck Institute for Meteorology as well as the World Climate Research Programme (WCRP). *Heike Oevermann* (born 1970) is architect and teaches heritage studies. Together with 13 colleagues she represents COST Action CA18204, a Europe-wide urban planning-related project, on which they report concerning the co-production of knowledge in the context of urban heritage research.

Chapter 9
Between *Parrhesia* and *Fake*: Scientific Responsibility Today

Rainer E. Zimmermann

> *Fama, malum qua non aliud velocius ullum:*
> *mobilitate viget virisque adquirit eundo,*
> *parva metu primo, mox sese attollit in auras*
> *ingrediturque solo et caput inter nubile condit.*
>
> *Virgil: Aeneis IV 174–177. ("Fama, an evil, swifter in the run*
> *than any other, / is strong by agility, / acquires strength in*
> *walking, / small at first out of fear, then grows quickly into the*
> *air, / strides along on the ground and hides its head between*
> *clouds." (My own English translation following the German*
> *translation by Johannes Götte, Wissenschaftliche*
> *Buchgesellschaft Darmstadt, 7th edition of the Artemis edition,*
> *Munich, Zurich, 1988, 143)).*

Abstract While the public discourse has recently been increasingly burdened by the dissemination of *fake news* and *alternative facts*, the scientific community is now also threatened by such influences. This must be confronted and countered decisively: Scientific discourse thrives on free speech, which is both the right and duty of serious scientists, and individual imagination is also essential. This does not mean, however, that anything can be disseminated at will, but imagination in a scientific context always means *exact* imagination, i.e., one that has to submit to the strict conditions of consistency while remaining connectable to the knowledge acquired so far.

R. E. Zimmermann (✉)
Clare Hall, Cambridge, UK

Institut für Design Science, München e.V., Germany
e-mail: pd00108@mail.lrz-muenchen.de

© The Author(s) 2022
H. A. Mieg (ed.), *The Responsibility of Science*, Studies in History and
Philosophy of Science 57, https://doi.org/10.1007/978-3-030-91597-1_9

I

A defining characteristic of the Greek *pólis*, a type of settlement that began to dominate the entire Mediterranean region for more than a millennium from about 750 B.C., is that there is political decision-making by majority vote after a debate in an assembly. In this assembly (called *ekklesía*), to which all adult, free men of the city-state belonged, the authoritative institution was free speech (*parrhesía*). In so far as this speech was free in the sense that any member of the assembly could stand up and express his view of a problem in question, it manifested the diversity of these views, which were in competition with each other. In his still authoritative book on "Violence and Harmony,"[1] Massimo Cacciari pointed out that this (literally "political") competition can only be valuable "as a search for the forms and modes in which harmony is created and revealed. The purpose of *agón* is to *aletheúein*, to produce and reveal harmony."[2] Cacciari continues: "And this 'harmony', this connection between *pólemos* and *stásis*, between external war and 'internal war', *is* the *pólis*—a structure that the Orient never knew and will never know."[3] The inner opposition of all members of the assembly (*stásis* thus) is revealed precisely through free speech, and at the same time it is dynamically stabilized and framed. Cacciari: "It is precisely by asserting my difference from the Other, my uniqueness, that I am *with* him—or rather, I *stand* and thus inevitably confront the one who in turn is facing me (stásis), and recognize myself in the confrontation as being *with* him."[4] (Incidentally, we recognize an existentialist figure of thought here.) Speech is admittedly subject to its own set of rules. Syntax and semantics correspond to each other, never losing sight of the desired harmony (we would say today: the metastable balance). If the rules are not observed (literally also the rules of grammar), *parrhesía* turns into *hubris:* The expression of this is *stásis*, and this is contrary to nature, and therefore violates the system of the polis and thus contradicts the fundamental principle of Greek ethics, the principle of *kátà physín* (what is according to nature). There is a revaluation of the terms. Again Cacciari: "In 'alógistos' daredevilry is called courage, caution is called sloth, moderation is called cowardice, sectarianism is considered more than the bonds of blood, and the oath is no longer taken in the name of divine law, but to break human laws."[5] Here Cacciari refers to the causes of the Peloponnesian War, as so vividly described by Thucydides. But the same applies to less spectacular incidents during this long-running war, such as the strange and

---

[1] Massimo Cacciari: Violence and Harmony. Hanser, Munich, Vienna, 1995 (Adelphi, Milan, 1994). For a more structural–theoretical overview see my essay: "Mesógios – Zur Struktur der Polis-Netzwerke," in: Richard Faber, Achim Lichtenberger (eds.), A Pluriverses Universum, Civilisations and Religions in the Ancient Mediterranean, Fink/Schöningh, Paderborn, 2015, 113–130.

[2] Cacciari, op. cit., 7.

[3] Ibid., 13.

[4] Ibid., 22.

[5] Ibid., 40.

mysterious Hermenic crime of the year 415, which Christian Mann has reported on in detail and which, incidentally, is of considerable relevance to the current political situation in Europe.[6]

It is no coincidence that free speech has developed its significance particularly strongly within fields where *communication in science* (we say today: the discourse of science) is concerned, following a concept of science as it was propagated—at the earliest half a millennium later, namely in 1115 AD—in connection with the founding of the first university in the modern sense, in the *commune di Bologna*.[7] Although this early medieval science is essentially authoritarian and derives its material from writings accepted as binding them to its entire method, which is initially called "scholastic," it is nevertheless based on those critical elements that were taken over from the Greek tradition, which, because of its origin in the late Eastern Roman Empire, is rather a mixture of ancient and Middle Eastern thought, also with North African influences: first of all, dealing with the effort to clarify what it actually is that is being talked about. Thus, it already points to (at least logical) argumentation and not just idle gossip. Thus, *lectio*, the commentary on reading, remains fundamental, always linked to a formal, grammatical analysis that must precede any reflection. But the comment immediately triggers a discussion. In this respect the text recedes behind the search for truth; the *lectio* is transformed into the *quaestio*, whose conclusions lead to the *determinatio*. And finally, the whole system even detaches itself from *lectio* and gains its own autonomy in *disputatio*.[8] But with that, since about the thirteenth century, long before the beginning of the Enlightenment era, there is no turning back as to its basis: Whoever disputes in a proper way is concerned with grasping *what is the case* and recognizing the consequences that follow. And as we know from history, in the future the fields of faith and knowledge will become increasingly divergent. And in the end, only scientific authorities are recognized in the field of science, but no longer traditional ones, neither those of the monarchy nor those of the Church. (Of course, we ourselves are only gradually awakening from the nineteenth century in this respect while being within the twenty-first century, as Walter Benjamin already knew, just as the era of the Enlightenment with its specific consequences really only came to its awakening in the nineteenth century). But after an intermezzo of increasing rationality that has now lasted for some two hundred years, this scientific paradigm is under dangerous attack. Representing the sciences, philosophy (understood by Hans Heinz Holz as the science of totality) in particular is today exposed to the attack of two tendencies, which have always been present (at least since Greek antiquity), but which have recently become widespread and considerably stronger in society: On the one hand, its (philosophy's) concept is inflationarily damaged because many people believe

[6] Christian Mann: The Demagogues and the People. On Political Communication in Athens in the fifth century B.C. Academy, Berlin, 2007, especially: 244–261.

[7] Cf. Rainer E. Zimmermann: Η ΝΕΑ ΠΟΛΥ. New City Concepts on the Way Home. LIT, Berlin, 2014.

[8] Jacques Le Goff: The Intellectuals in the Middle Ages. dtv/Klett-Cotta, Munich, 1993 (1986), 99th (par.) (Originally after the second edition of 1984 at du Seuil, Paris.)

that ordinary thinking is already philosophizing. Not to speak of the space-consuming and annoying habit of speaking of so-called "corporate philosophy" or the "philosophy of a product" that is nothing but a banal "purpose" for the making of profit. This is accompanied not only by an explicit *trivialization* (in the sense that immersion in the professional depth is avoided at all costs and the easy apparent replaces the difficult actual), but also by an increasing *professionalization* (in the sense that the opinion prevails more and more that the activity of philosophizing can be carried out by any person without great difficulty, in such a way that a regular income can be drawn from it, because the possibility of a real, concrete and insofar helpful consultation of fellow humans is apparently imminent). This double misunderstanding is not only prevalent in everyday life today, but also affects those who should know better by means of their profession, i.e., numerous scientists as well as artists—and in fact several philosophers themselves. While it is obvious that it is the currently precarious job situation, especially in the academic middle class, that provides the incentive to look for a replacement, this cannot really excuse either the essential trivialization of the activity or its shift to appearances. In the case of a space engineer or a theoretical physicist, it would be difficult to assume that their activity could be carried out by people who have not undergone the usual training for this activity. Incidentally, one would not dare to appear as a cook or baker or carpenter without the appropriate training. (The few who sometimes do so are rightly called impostors.) But the humanities and philosophy in particular have always been increasingly exposed to this arrogance.[9] Unfortunately, today we are increasingly encountering comments, be they in the form of individual scientific lectures, contributions to conferences, or written publications, in which we feel as Loriot once did in the well-known sketch about the art whistler, who shows up for an interview, but who clearly does not master his profession at all. Then Loriot, in the shape of one of his famous cartoon characters, says: "But, Sir! There's no art to it!"[10]

On the other hand, the phenomena of targeted fake news and *alternative facts,* which have only recently been addressed to a large extent, are increasingly relativizing, questioning and damaging the very concept of science itself, which in turn underlies the concept of philosophy, so that orientation in the midst of the concrete world is increasingly lacking, and central criteria of thought that have made possible in the first place the progress of thought over the last three centuries (at least) are in danger of being lost. Obviously, the first tendency is directly related to the second: For political interests in general were already (in the past) such that people did not shy away from misinterpreting facts in their own favor and spreading their opinions under false assumptions or simply denying the existence of facts. It is indeed

---

[9] That's why Freud once signed a birthday card to Einstein saying, "You lucky one." In response to Einstein's question, he explains that no one would dare to have a say in Einstein's work, whereas everyone feels called upon to make Freud's work their own field of expertise—a topic for the regulars' table.

[10] "Aber mein Herr, das ist doch keine Kunst!" Episode 10 of "Loriot" (TV series Radio Bremen), 1976–1978. Available at https://www.youtube.com/watch?v=pC2tXEOjLLk (20.03.2019).

characteristic of the human exercise of power, or of the striving for this power, that it is mainly about the *instrumentalization of fellow human beings*: By believing what is communicated to them by the world-readers, people serve the interests of those whom the world-readers serve. Jean-Paul Sartre already knew that everyone pursues their own interests. But because the others are doing the same thing at the same time, these interests overlap and interfere with each other, and what emerges in the end as a result, sometimes later called a "historical event," is essentially *counterfinal*, i.e., something that none of the participants really wanted, but which was spontaneously created in the course of the overlapping. In this sense, as Karl Marx once so aptly put it, "[t]he people make their own history, but they do not make it of their own free will, not under circumstances chosen by themselves, but under circumstances that are found, given and handed down directly. The tradition of all dead generations weighs like an alp on the brains of the living. And when they seem to be busy transforming themselves and the things, creating something unprecedented, it is precisely in such periods of revolutionary crisis that they anxiously invoke the spirits of the past at their service, borrowing from them names, battle slogans, costumes, in order to perform the new world history scene in this time-honored disguise and with this borrowed language."[11] It is precisely the history published and later disseminated in schools (as a subject, thus understood as a science of history) that has in the past often made a considerable contribution to interpreting historical events from the point of view of particular interests, namely depending on the specific interests of those who "wrote" this particular history, largely independently of actual events. The recent qualitative difference is basically that this tendency towards pragmatic instrumentalization has long since transcended the realms of immediate "domination" and found its way into the ordinary everyday life of social institutions, where it can be multiplied *en masse* since the arrival of the "new media": There have always been people with strange views of the world. But now they can get together with like-minded people and make their ideas known and widely disseminated. People used to say, "X is a weird guy; What he thinks…"; And one would either avoid or ridicule such persons. Today it is said: "X spreads a strange opinion on the Internet and has numerous followers; So there must be something to this." At the university, they used to say, "X is a crackpot. His methodological approach is not up to standard." Today, it is often said: "X represents a special view and not the school opinion. And differences of opinion always exist within the framework of diversity." In the end, it is therefore no longer possible to differentiate sufficiently between the publication of a scientific paper in an established, peer-reviewed journal versus a self-administered *online text* on the Internet. The latter at least gives the impression of seriousness. However, it has not been adequately verified by specially designated *reviewers* (*referees*). Scientifically or generally ideologically, one is reminded of the story going back to Bertrand Russell, in which a practicing *tea-ist* (sic) claims that between Earth and Mars a teapot is circling in

---

[11] Karl Marx: The Eighteenth Brumaire of Napoléon Bonaparte. In: Karl Marx, Friedrich Engels: Werke (MEW), German Edition, Dietz, Berlin (GDR), 1972, Volume 8, 111–207, here: 115.

space, but it is so small that it cannot be detected by telescopes. The tea-ist insists on this view as long as no one can prove him wrong. In contrast, the sceptic, the *A-tea-ist*, points out that it is not *his* task to refute curious assertions, but that the burden of proof lies solely with those who make these assertions. The instruments for this procedure, namely the propositions of the sufficient ground and of "Ockham's razor," seem to be rather forgotten today.[12]

And what is true on a small scale and for everyday life, both scientific and non-scientific, is also true for the larger social perspective: in this way, the constitution-ality of human beings, so plausibly described by Marx, determines politics (i.e., daily politics as well as the major geopolitical constellations), but at the same time also the behavior of people in everyday life, which is situated below the political horizon. On the one hand, there is the global outbreak of "overriding" interests, regardless of whether some two and a half thousand years ago a battle for hegemony in the Mediterranean broke out between Athens and Sparta because Sparta feared a reduction in power; a thirty-year war was fought in the seventeenth century because the Bavarian Elector was jealous of the Habsburgs' claim to power; or whether the nations "sleepwalked" into the First World War because each was concerned about its "place in the sun." In the first case, the claiming of due respect is put forward in order to enforce claims to hegemony that cannot really be justified. In the second case, the fighting in favor of the true Christian religion is pretended. In the third case, it is allegedly a matter of maintaining a balance that was once constructed at the time of the Congress of Vienna. Nevertheless, in all three cases it is in fact a matter of striving for concrete hegemony as a basic prerequisite for one's own fur-ther growth. On the other hand, however, this state of affairs continues to be reflected in the regional, local, and "microscopic" interests of social groups and individuals: even beyond the politically relevant institutions, the main issue is one's own "hege-mony," no matter how small the corresponding sphere of rule, no matter how insig-nificant its object. And it is always the individual discourse that constitutes the conditions of validity. For an appropriate illustration I give an example in the following.

II

For this purpose we use the recently published study by Frank Westerman about a catastrophe during 1986 in the so-called "Valley of Death" in western Cameroon,[13] which can be used to illustrate the facts mentioned here: On the night of 21 to 22

---

[12] The same argument can be used when it comes to the existence of pink elephants that disappear whenever one tries to look at them or, conversely, the Flying Spaghetti Monster—originally con-cocted to lampoon the equivalence accorded by the Kansas State Board of Education to teaching pseudo-scientific "intelligent design" as an alternative to Darwinian evolution.

[13] Frank Westerman: The Valley of Death. A Catastrophe and its Invention. Left, Berlin, 2018 (De Bezige Bij, Amsterdam, 2013).

August, about 2000 people and numerous animals, including birds and insects, died in a very short period of time in the Nyos Valley, about 300 km northwest of the capital Yaoundé. Hundreds of injured were taken to a hospital in the nearby town of Wum. Reported symptoms included "blister-like ulcers" and "suffocation and choking." According to the current state of research, the disaster was most likely caused by a specific constellation of the water body immediately after the Nyos Lake incident, formerly called Lwi Lake ("The Good Lake"), but since called "The Angry Lake." This is because western Cameroon belongs to the famous "breaking point" where Africa "broke off" about 100 million years ago in the course of continental drift from South America. The fracture line of instability is also called the "Cameroon Line," which extends from some offshore islands over Mount Cameroon, an active volcano 4000 m high, further north, where there is a chain of volcanic lakes, which form mostly circular *maars*, i.e., volcanic craters formed by explosions when groundwater meets liquid lava. In addition to Lake Nyos, there are two other lakes (Lake Monoun, also in Cameroon, and Lake Kivu in the area between Rwanda and the Democratic Republic of the Congo, formerly Zaire) in which carbon dioxide is dissolved to saturation point. Because of the thermodynamic conditions of temperature and pressure, deep water can store far more carbon dioxide than surface water. Subterranean magma chambers ensure the permanent supply of further carbon dioxide. A spontaneous perturbation of the system, such as a landslide or small earthquake, can lead to explosive outgassing. Because carbon dioxide is heavier than air, it then flows very quickly (with outflow speed in the range of hundreds of km/h) near the ground into all surrounding lowlands. Carbon dioxide is odorless and invisible and, at concentrations greater than 8% of the air we breathe is almost certainly fatal. A comparable outgassing had already occurred at Lake Monoun in 1984, costing 37 people their lives. In 1986, Lake Nyos suddenly released almost two million tonnes of carbon dioxide. The whole area was declared a military exclusion zone, and even today many victims and their descendants live in camps rather than in newly established settlements elsewhere. Since 2001 a French team has been trying to carry out a degassing project on the lake.

However, Westerman focuses less on the catastrophe itself than on the various discourses on it over the years: One can distinguish between a scientific discourse, a political discourse, a religious discourse, and a mythical discourse. The first, scientific discourse, is essentially determined by the competition between two schools, whose central protagonists are Haroun Tazieff (Paris) on the one hand and Haraldur Sigurdsson (Reykjavik) on the other. There is general agreement that the deaths are a consequence of the effect of carbon dioxide. The exact cause of the volcanic activity is, however, disputed. Sigurdsson understands it as the spontaneous emergence of instability. Both protagonists draw on their earlier experiences. On the event two years prior at Lake Monoun, Sigurdsson had written an essay propagating his theory, which he could not place in either of the leading journals "Science" or "Nature" and which he had therefore submitted to another, less prestigious, journal, but which took a long time to reach publication. This ultimately led to the fact that Sigurdsson, who was on Sumbawa in Indonesia during the Nyos catastrophe, where he was investigating the Tambora volcano, was not able to join the investigation of the more recent case in a timely manner, nor did he have anything in writing regarding the earlier incident. Hence, Tazieff was the first and "loudest" person to advance with his own theory and to announce his result even before leaving Paris. The scientists who subsequently arrived in the disaster area from all over the world were consistently divided between the two camps. But with the arrival of the US team, supporters of Sigurdsson's view finally gained the upper hand. The scientific competition was also reflected in the media: Reuters and AP stood against the French AFP, "National Geographic" magazine against the French GEO. In their respective announcements the competing view was simply omitted. At the large Nyos Conference held in Yaoundé in 1987, which was attended by 86 experts from 35 countries, 46 of them from Cameroon, the final communiqué was based entirely on the Anglo-Saxon-dominated view of Sigurdsson. This led to an angry and spectacular performance by Tazieff, who did not want to finish his counter speech and was therefore cut off the microphone, whereupon he demonstratively left the conference hall (incidentally, celebrated as a hero by many young participants in the conference, because his public opposition was very positively evaluated by the inhabitants of an extremely rigid dictatorship). Tazieff was also unable to assert himself at the subsequent UNESCO conference in Paris, whereupon he boycotted all further conferences on the subject and declined to participate.

The second, political discourse is closely interwoven with the first, because Paul Biya, President of Cameroon since 1982, often expected to be pushed out of office by the former colonial power France. He also had to endure many years of conflict with Nigeria and internal tensions resulting from the original division of Cameroon into an English-speaking part (which later became Nigeria) and a French-speaking part (Cameroon in the narrower sense). Nevertheless, English-speaking inhabitants had remained in Cameroon. The disaster area in question is one of the predominantly Anglophone regions. The ratio between the French-speaking majority and the English-speaking minority is about 80:20, and it is deemed quite likely (but difficult to prove) that Biya himself was involved in spreading such rumors, which linked the disaster in the Nyos valley to a secret weapons test by a foreign nation. Partly the French were accused of having carried out poison gas experiments, partly the Israelis of having tested a new bomb, perhaps a neutron bomb. But the USA was also accused in one way or another. Should Biya himself have contributed to these rumors, the atmosphere of mistrust was soon to backfire. In 1990, a vaccination program for girls and women between the ages of 14 and 30 was set up in the English-speaking part of Cameroon. The director of the Catholic Augustine College in Nso, Father Fonteh, publicly rebuffed this program, finding it suspicious that only specific groups of women should be vaccinated. Consequently, the vaccination program failed. Father Fonteh took a hypodermic syringe for analysis at a clinic run by nuns in Shisong. Shortly afterwards, he was beaten to death in his own home in a murder that was never solved. It subsequently transpired that the supposed vaccine was in fact a hormone preparation that sterilized women. Hence, in retrospect, the Nyos catastrophe was interpreted as a measure to decimate the remaining English-speaking population.

The third, religious discourse originated from the situation of Christian missions in the area: Westerman impressively describes the arrival of various priests in the disaster area. He describes the actual (yet almost unbelievable) meeting like a symbolic ritual: "Three white men, preachers of a non-African religion, enter the valley of death simultaneously, but independently of each other, as the first to come. One enters from the east. The other one descends from the sky [by helicopter]. The third comes down from the southern hills."[14] One of them, Father Jaap, says the first thing: "Isn't this Satan's work?"[15] The mixture of Christian missionary and late colonialism could not be put more succinctly: "They [the missionaries] bring a mythical narrative that is supposed to take the place of all other mythical narratives. It is about the chosen and the damned, about prophets and angels, about God and the devil. What they proclaim is written in a centuries-old book full of guidelines for life, from the cradle to the grave, and a reward for those who submit to them: the promise of eternal life."[16] On the other hand, it is also justified to note that the Christian missions are so popular because, especially after the Nyos disaster, they are the only ones to care for the victims in any way, even in a common ecumenical effort that includes Muslim communities. Of course, soon the White US evangelists, who represent a part of the mission, also enter the debate about guilt and are insofar exposed to suspicion. But even these rather mild, though very committed priests, in the end simply retreat to the infinite wisdom of the Lord. In this way, although they spread a mythical narrative that sees itself as a kind of "master narrative" that cancels out all other narratives, they essentially belong to the fourth, the mythical discourse.

The latter is based on the network of legends that goes back to the founding myth of the local tribes. An enlightening overview of the complexity and the overall context of intra-African migrations, especially from the seventeenth century onwards, is provided by the still significant historical work of Joseph Ki-Zerbo from 1978,[17] in which the volcanic lakes of the southwest, which are often regarded as the abodes of the spirits of the dead—a form of underworld equivalent—play a central role. It is obvious that past catastrophes of the kind mentioned have traditionally been equated with the intervention of those spirits or even gods hidden in the depths, and will certainly continue to be so.

Westerman describes very vividly the effect of the narration that has been handed down. On the one hand, he emphasizes a socialization characteristic that can be regarded as universal: "Narrations can so easily settle into reality that they become part of it. [...] Everyone in this world raises his children with food, drink and fairy tales. [...] / [...] When it comes to questions of life, the majority of the world's

---

[14] Ibid., 113.

[15] Ibid. 114.

[16] Ibid. 122.

[17] Joseph Ki-Zerbo: The History of Black Africa. Hammer, Wuppertal, 2nd edition 1981 (1979). Originally: Histoire de l'Afrique Noire, Hatier, Paris, 1978; for Cameroon, see in particular: 311–313.

population prefers fiction to fact."[18] He sees narrative as a "regrinding of raw reality," and continues: "We all [always] felt a tendency to recognize connections that perhaps did not exist in reality, but which, by naming them, acquired a certain persuasive power";[19] And later says: "Human curiosity is not content with incompleteness, inconsistency or incomprehensibility. If there's no other way, we'll add what's missing."[20] In this context it is almost trivial to note that the origins of the myth are essentially based on the fear of death, i.e., the fear of the finiteness of human life.[21] Westerman quotes Malinowski: "The myth does not explain, [it] justifies."[22] In the case of the disaster in Cameroon, however, the logic of the myth did not work: According to the polar view of tradition, the premature death of the victims was something unnatural, i.e., evil. Therefore, it could only be a punishment or revenge of the ancestors. (A figure, by the way, which indirectly also corresponded to the Christian understanding of guilt). On the other hand, the large number of victims was something that was completely disproportionate to any possible guilt and thus did not fit into the traditional understanding of the myth.[23]

From this episode one can see quite clearly the problem we have to deal with *today:* At that time (i.e., especially in the 1980s) there were different discourses via which one could report on a phenomenon. Basically, everyone chose the discourse that was close enough. There was little overlap, which sometimes served the purpose of interest-based mystification, and sometimes came from a random constellation that contributed significantly to the lack of clarity. But there was no real mixing, because the demarcation between the discourses was largely adhered to and observed. Of course, systematic and methodological mistakes were made from time to time: For example, after the catastrophe in the Nyos valley, the focus was initially on questioning witnesses.

However, only later was it revealed that the survey was not conducted in the specific tribal language but in the pidgin English spoken in southwestern Cameroon (one could say: the *lingua franca* of the region). However, this is a language that simplifies matters very much, so that the color of the lake after the disaster was described in a rather general way. (The surface of the water was called "red," but in pidgin English this is equivalent to "yellow" or even "blue." Westerman has verified these facts on site. Sam Freeth, the British expedition leader was the first to point out this aspect).[24] Likewise, it was not possible to distinguish precisely between "smell" and "taste." It goes without saying that both the color of the lake after the event in question and the associated smell and/or taste were or should have been able to provide essential information about the chemical substances involved.

---

[18] Westerman, op. cit., 24 sq.

[19] Ibid., 150 sq.

[20] Ibid., 176.

[21] Ibid., 252. (par.)

[22] Ibid., 302.

[23] Ibid., 286 sq. (par.)

[24] Ibid., 80 sq. (par.)

So the important aspect in this context is the *diversity of discourses*, which I have presented here in a very simplified form. Michel Foucault and his collaborators had already carried out a similar investigation—into the influence of various discourses that overlapped only slightly—when researching in 1973 the case of Jacques Rivière.[25] Although this was not a collective catastrophe but an individual case of murder in a rural family in 1835, the influence of the tension that existed between the different types of interpretation to which the event in question was subjected under the impact of the juxtaposed discourses were also clearly evident in this example.

But what is different today? The manifold discursive perspectives have always existed. Essentially, however, there are now two differences: First, the implicit weighting of discourses has largely disappeared. Due to exaggerated politeness or even misunderstood "political correctness," the various discourses are now considered equal and therefore *indifferent* (in a double sense of the word). On the other hand, all discourses in the media, especially in the new media, are equally widespread and accessible everywhere at any time. This means that rumors spread much faster than in the times of Virgil. This easily leads to confusion and unjustified relativization, and all this adds to the confusion. Strange assertions or just nonsense can easily hold their own from now on. The media presence of nonsense therefore gives the uninformed the impression of justified commitment. However, this tendency had its origin less in philosophy or the sciences, but rather in the structure of public discourse. This structure actually started with the reserved politeness of knowledgeable people who did not want to reveal themselves as knowing in order not to offend their fellow humans and to avoid distinguishing themselves as an alleged "elite." When in TV talkshows semi-prominent actresses spread their "health-giving stones" on the table, people only nodded amiably with a serious look; no criticism was expressed—on the contrary: criticism was explicitly frowned upon. For the time being, it ended with a levelling of all views, also in the scientific field. By the way, supported by the increasing commercial professionalization, which is mainly subject to the criteria of advertising (i.e., propaganda). For if the number of "clicks" is directly or indirectly rewarded by money, those offers that transport nonsense, up to and including explicit perversion, will always prevail. It is not without reason that media networks such as "Facebook" secretly employ numerous employees who have the task of checking uploaded content and deleting it if necessary.

The final result makes a mockery of the principles of the Enlightenment. My Munich colleague Silke Järvenpää quoted US philosopher Laurie L. Calhoun as an example at a recent conference on the subject in Berlin.[26] The science journalist

---

[25] Michel Foucault (Ed.): The Rivière Case. Materials on the Relationship between Psychiatry and Criminal Justice. German edition at Suhrkamp, Frankfurt am Main, 1975 (Gallimard/Julliard, Paris, 1973)

[26] Silke Järvenpää: 'Alternative Facts' and 'Fake News': Applied Cultural Studies? In: Rational and Irrational Discourses in the Age of Digitalization. HTW Berlin, September 21, 2017. Published as: Alternative Facts and Fake News: Cultural Studies' Illegitimate Brainchildren, in: Ralph-Miklas Dobler, Daniel Ittstein (Eds.), Fake Interdisziplinär. UVK, Konstanz, 2019, 71–83.

Helen Pluckrose refers to Calhoun as follows: "Calhoun […] redefines the building blocks of scientific methodology as a contemporary form of magical thinking. A colleague recalls a discussion: 'When I had occasion to ask her whether or not it was a fact that giraffes are taller than ants, she replied that it was not a fact, but rather an article of religious faith in our culture.'"[27]

Incidentally, Calhoun is also a *research fellow* at the Independent Institute, an apparently non-profit educational institution in Oakland, California. As far as can be seen, she does not hold a professorship, as is often rumored. In return, however, for the sake of fairness, it must also be said that Helen Pluckrose in her contribution does indeed make justified criticism of so-called "postmodernism," although she herself takes a very woodcut-like and simplistic approach.

III

I conclude and summarize: While the public discourse has recently been increasingly burdened by the dissemination of fake news and *alternative facts*, the academic world is now also threatened by these influences, which aim to contaminate and discredit research and teaching through unscientific claims, esoteric thinking, politically motivated ideologies and misunderstood political correctness. This tendency is supported by a publication industry that takes a pseudo-scientific approach and publishes all kinds of things for a fee yet without verification. Only recently, it has been established that the number of scientists taking advantage of such opportunities is already in the thousands in Germany alone. This must be stopped decisively: Scientific discourse thrives on free speech, which is both the right and duty of serious scientists, and individual imagination is also essential. This does not mean, however, that anything can be disseminated at will: imagination in a scientific context always means *exact* imagination, i.e., one that must submit to the strict conditions of consistency while remaining connectable to the knowledge acquired so far. The instruments for checking consistency are already in place and basically sufficient, but may need to be rethought in the light of changes in the media landscape.

---

[27] Helen Pluckrose: How French 'Intellectuals' Ruined the West: Postmodernism and its Impact, Explained. https://areomagazine.com/2017/03/27/how-french-intellectuals-ruined-the-west-postmodernism-and-its-impact-explained/ (10.03.2019)

Chapter 10
The Responsibility of Science for Guaranteeing Human Rights in the Fight Against Human Degradation, Racism and Anti-Semitism

Klaus Fuchs-Kittowski

Abstract This article is an appeal in the form of ten questions or theses, which as a whole outline the responsibility of science. 1) The humanistic mission of science, respecting human rights, against reductionism; 2) Against reification and the degradation of the living; 3) Information creation: An essential category for model and theory development and as a general methodological guiding principle! 4) The need for education and training against racism: The responsibility of science in the fight against anti-Semitism; 5) Deadly science: The demand for the destruction of life unworthy of life was "scientifically" justified; 6) Religious traditions as a cause of anti-Semitism? 7) Causes of anti-humanism, racism, anti-Semitism, and neo-Nazism in the contemporary world of work; A deeper "perception" of life and human beings is also necessary in the economy! 8) People must be able to recognize the meaning and purpose of their existence 9) Peace must be secured! 10) Science can follow its humanistic obligation to serve life, to serve mankind.

Jews are targets of hate, including in Berlin! The "Reach Out" victim advisory center reports an increased number of racist or anti-Semitic attacks on Jewish citizens in Berlin. In 2018, 309 attacks were documented. This is 46 acts of violence and massive threats more than in 2017.[1] Those affected by hate crimes can experience long-term (even lifelong!) traumas and disorders. The social climate has become much rougher in recent years. Why do adult men use violence against children and adolescents for racist reasons? In this matter, our society, including Berlin, has a serious problem!

---

[1] Annika Leister, Zielscheibe des Hasses, Berliner Zeitung No. 56, 7/8 03, 2019, p. 15.

K. Fuchs-Kittowski (✉)
Leibniz-Sozietät der Wissenschaften zu Berlin, Berlin, Germany

© The Author(s) 2022
H. A. Mieg (ed.), *The Responsibility of Science*, Studies in History and Philosophy of Science 57, https://doi.org/10.1007/978-3-030-91597-1_10

The Humanistic Mission of Science, Respecting Human Rights, Against Reductionism

J. D. Bernal, the founder of Science Studies, claimed that science has a social function: "Fortunately science has a third and more important function. It is the chief agent of change in society; at first, unconsciously as technical change, paving the way to economic and social changes, and, latterly, as a more conscious and direct motive for social change itself."[2]

Here it should be made clear that, as J. D. Bernal points out, this conscious and direct motive of science to contribute to social change requires that science fulfil its humanistic mission. This means, however, that through its results science contributes to the guarantee of human rights, that it assumes its responsibility: to secure the truth of its statements and to realize an application of the truth that serves life and man, and thus also sees its responsibility in promoting, not hindering, a deep "perception" of life and man, so that nature and man are recognized and acknowledged in their specificity and value.

If, as the long-standing IFIP (International Federation for Information Processing) President Klaus Brunnstein repeatedly emphasized, it is assumed that there are not only individual but also social and international human rights, this means, for example, for the work of computer scientists, standing up for data protection, as an individual human right; for work and organizational design that promotes personality development, as a social human right; as well as for a life of peace, as an international and first human right. When we work to guarantee human rights, the inviolability of the dignity of every human being is paramount, and with it the fight against any form of degradation of the living, against racism and anti-Semitism.

J. D. Bernal recognized that society can only achieve its ambitious goals with the help of science, but that the social effectiveness of science is largely determined by the introduction and mastery of modern methods and techniques of research, and also by the organization and management of social processes. For the progress of knowledge, the traceability of complex processes and structures to the underlying elementary processes and structures is a decisive prerequisite. However, it is important to realize that one must not stop at reduction, because it is too limited for the realization of the whole. Today, a special responsibility of science arises in particular from the fact that a one-sided, reductionist, scientific-technical culture obviously leads to a loss of perception of life and man. Under certain social conditions this can in turn be abused by racists and other anti-humanists.

---

[2] John Desmond Bernal, The Social Function of Science (1939) Faber & Faber, p. 383.

Against Reification and the Degradation of the Living

Life, with its unique, highly complex structure, is exposed to many dangers and this, as many authors point out, not only through changes in external conditions such as the greenhouse effect, but also and perhaps even more so through this loss of perception of a reductionist scientific and technical culture towards the living and towards man. It is not the discovery of nuclear fission or DNA, and now the decoding of the human genome, nor the development of the computer and, at present, the global digital networks—the Internet and the Internet of Things—that constitute this threat to our world. Rather, it is based on the fact of a far-reaching reification and degradation of the living, in that everything is now only regarded as a usable resource and treated accordingly. This ruthless urge to exploit, by which every new scientific hypothesis is immediately put to the test of its potentially profitable applications, characterizes the current zeitgeist to a large extent.

It is a legitimate aim of bio-medical research to uncover the causes of diseases that are still incurable today, such as Alzheimer's, cancer, and Parkinson's, and to seek ways of curing them. An intervention in this complex happening of life-processes should not be demonized as hubris. However, plans to improve human health as a whole, developed by misguided ambitions of scientists, or the hasty introduction of new products driven by the greed of some companies, must be firmly rejected as a violation of human rights. Here, in fact, the contempt for man, the degradation of all living things, the absolute necessity of exploitation under the prevailing economic forces, comes to the fore.

It is the responsibility of science and scientists to ensure that important scientific-technical developments—currently especially in computer science and biology—are not misused to underestimate or even disregard human beings in their complexity, sensitivity, uniqueness, individuality, etc.

It is the basic attitude of a reductive, primitive mechanistic materialism, which has been spread with the great successes in modern science, especially in biology and computer science, that provides the breeding ground for religious fundamentalist movements. If the mind is generally denied, which identifies with information processing and attributes this to signal processing; or syntactic information processing is reduced when, in the name of modern science, it can be generally explained that man and computer are identical—'it is only hardware or wetware'; when, as the latest finding of science, the identity of mind and brain, the reduction of the mind to neuronal connections[3] or connections of small robots,[4] is proclaimed everywhere, then one should not be surprised that with a widespread lack of prospects for humans, a counter-reaction is thus triggered so that—as is becoming apparent even

---

[3] Francis Crick, The Astonishing Hypothesis: The Scientific Search for the Soul. Scribner, 1994.

[4] Daniel C. Dennett, Sweet Dreams: Philosophical Obstacles to a Science of Consciousness. MIT Press, 2006.

in the rich countries—one wishes for the "intelligent designer."[5] It is not entirely surprising, then, that this is becoming a mass movement even in parts of Europe, or that one is turning to other fundamentalist groups and to a body of ideas which also promotes racism.

The reduction of human beings to animals, and the inferiority of parts of humanity in biological and spiritual terms that this claimed, was one of the important ideological prerequisites for both world wars. The reduction of man to the machine; the currently widespread postulate that automata could even become better people and a post-biological age could dawn; human society could be replaced by an automaton society, as postulated by the MIT robotics/artificial intelligence researcher Hans Moravec in his book Mind Children[6], thereby anticipating the complete destruction of humanity. Even such false ideas have power, as J. Weizenbaum[7] and Benno Müller-Hill[8] never tired of reminding us, since they contribute to the degradation of human beings and thus to racism and anti-Semitism.

Information Creation: An Essential Category for Model and Theory Development and as a General Methodological Guiding Principle!

Reductionism in science as an ideological attitude can and must be countered by working out the specifics of the living, especially of the living in relation to the dead; especially of the human in relation to the technical automaton, the so-called autonomous robot. Joseph Weizenbaum asks Hans Moravec whether he can really assume that one can transfer the truly human—e.g., a smile of a young mother to her child—to robots.[9]

The learning automaton, including the those under development for so-called autonomous vehicles, receives its information and its value system from outside. In the origin of life, as Manfred Eigen[10] demonstrated with his Darwinian theory of the origin of life, the information and the value system must originate internally.

[5] Ronald L. Numbers, The Creationists: From Scientific Creationism to Intelligent Design, University of California Press, 1993, Harvard University Press, 2006.

[6] Hans Moravec, Mind Children: The Future of Robot and Human Intelligence, Harvard University Press, 1990.

[7] Joseph Weizenbaum, Computer Power and Human Reason: From Judgment to Calculation. W.H. Freeman and Company, 1976.

[8] Benno Müller-Hill, Die Philosophie und das Lebendige. Campus Verlag, 1981.

[9] Klaus Fuchs-Kittowski, Bodo Wenzlaff, Probleme der theoretischen und praktischen Beherrschung moderner Informations- und Kommunikationstechnologien. – In: Deutsche Zeitschrift für Philosophie (Berlin). 35(1987)6, S. 502–511.

[10] Manfred Eigen, Molekulare Selbstorganisation und Evolution, in: Joachim-Hermann Scharf (Hrsg): Informatik, Nova Acta Leopoldina, Johann Ambrosius Barth: Leipzig 1972, S. 171–223.

The category of information creation proved to be essential for understanding the origin of life, for model and theory formation in the border area between physics, chemistry, and biology. Wherever functions need to be newly created and organized, new information and evaluations are required. Therefore, the category of information creation is as essential for the understanding of phylo- and ontogenesis, as well as for model and theory formation in the border area between computer (software) and the human mind, as well as automat-supported information systems (application systems) and creative-learning social organizations. It is part of the responsibility of science, especially of a theory of biology as well as a theory of computer science, to bring this specificity of the living and of man to bear, because this is the only way to protect decisively against the degradation of man to the automaton and thus also against further forms of discrimination.

There are, also scientific-theoretical and methodological implications of the concept of creativity: the creation of information has gained in importance for almost all areas of scientific interest. In particular, there is methodological evidence for safer navigation between the Scylla of gross reductionism, (inspired by 19th-century physics) and in the 20th century by the "mind–brain identity" (neurophilosophy) of connectionist AI research and the Charybdis of dualism (inspired by the vitalism of 19th-century Romanticism and in the 20th century by the functionalist body–mind / hardware–software duality of cognitivist AI research.

The basis for post-humanistic and other anti-humanistic concepts is the reduction of the human being to an information system and the reduction of information to its syntactic structure, in accordance with the information processing approach of classical AI research. The central idea of creativity—namely the origin of information in the living, in creative thinking and in an evolving, living social organization—leads to an understanding of human–computer interaction as a coupling of machine (syntactic) information processing with the creatively active human being capable of semantic information processing. Thus, the goal of automation is not superautomation, i.e., the complete replacement of humans, but the meaningful coupling of the specific abilities of computers and humans. This means that anti-human ideas are also losing their theoretical and practical ground.

The Need for Education and Training Against Racism: The Responsibility of Science in the Fight Against Anti-Semitism

The fight against racism must be a major concern in the debate about the responsibility of science and scientists. It was argued that there is no need for scientific arguments against racism, since a humanist must be against all forms of racism from the outset, regardless of any scientific evidence. Even if this is correct in principle, because ethical values do not come from natural science but from experiences of the social life of human beings, the scientific insight that there is no "cultural gene," for example, can be useful, as postulated by some molecular biologists and philosophers under the assumption of a strict genetic determinism. The genes have nothing to do with what is meant by being human.[11]

We understand humans as a bio-psycho-social unit, a biological–psychological and predominantly social being. The basic physical and mental abilities of humans are determined by their genes. In evolution, selection has been based in particular on skin color, speed of movement and perception of the environment, but not on those characteristics that would allow one person to be considered inferior to another; So—as previously stated—not about one being human. This is an expression of the whole person.

Racism is a false, extremely dangerous ideology! But it would therefore be naive to simply dismiss it as an ideology and not to argue scientifically against it. Especially now that findings from human genome analysis show that "Genetic differences exist between ethnic groups, which relate to external characteristics and parameters of metabolism. They have nothing to do with 'being human', but could serve to camouflage misanthropic statements by a racism of any kind, a division of people into good versus bad races, falsifications and distortions of science."[12] Society must ensure that its members consider all people (including the disabled) as 'equal', regardless of genetic differences between individuals and groups.

The central ethical question posed by modern biotechnology and modern information technologies in a new quality is that of the constructability and replaceability of human beings.

On the one hand, science bears the responsibility for obtaining genuinely true statements about the world around us and about ourselves, but also for ensuring that this truth is applied for the benefit of mankind. It is clear that this responsibility thus goes much further than gaining methodologically secured knowledge—that it ultimately has our own existence as human beings and our ability to make ethical decisions as a problem. There are enough examples in recent history of biological facts (for there are indeed human races, not only animal species and plant varieties) being deliberately and erroneously misunderstood and misinterpreted (racism).

In memory of the 6 million European Jews who were killed in the concentration and extermination camps of German fascism, it should be one of the most important tasks of science, especially of German scientists, to fight anti-Semitism in all its forms, to contribute to education against hatred and violence and anti-Semitism, and above all to help to overcome the social causes that repeatedly produce such an inhuman ideology. We hear that every year thousands of Jews leave France and emigrate to Israel. Reporting on this issue, "Der Spiegel" commented: "Anti-Semitic attacks have increased by 74 percent in 2018, a surely unbelievable rise. 500

[11] Klaus Fuchs-Kittowski, Marlene Fuchs-Kittowski, Hans-Alfred Rosenthal, Biologisches und Soziales im menschlichen Verhalten. In: Deutsche Zeitschrift für Philosophie, Heft 7, 1983, S. 812–824.

[12] Benno Müller-Hill, Die Gefahr der Eugenik. – In: Was wissen wir, wenn wir das menschliche Genom kennen? Hrsg. v. L. Honnefelder u. P. Propping. DuMont 2001, pp. 218–219 (translated).

incidents were recorded, up from 311 the year before."[13] In Germany, as in many other countries, the number of attacks on Jews has increased, as in Austria and Switzerland.

What are the reasons for a development that, based on historical experience, was hardly considered possible? After all, we do not have an economic crisis in the countries mentioned, comparable to that of the 1920s, for which a scapegoat was sought and found. It is not possible here to go into the multitude of possible prejudices, dating back to the Middle Ages, which were the breeding ground for anti-Semitism, against which there has always been a fight. Active work is also ongoing to overcome the various causes that stoked such prejudices in the past. At present, one of the obvious causes is the tense relationship between the Palestinians and Israel. Above all, the old mechanism of demagogues is being used again. Real contradictions are addressed, no (or inappropriate) solutions are offered and a scapegoat is sought for grievances that have still not been eliminated.

Deadly Science: The Demand for the Destruction of Life "Unworthy of Life" Was "Scientifically" Justified

A number of German researchers were deeply involved in the crimes of the German fascists, directly or indirectly promoting them through their scientific work. "What that meant sounds unimaginable today: Scientists of the Kaiser-Wilhelm-Gesellschaft (KWG-Biologen) ordered the eyes of murdered people directly from the concentration camp doctor Josef Mengele in Auschwitz." […] "They were pairs of eyes of twins, of Sinti and Roma," reports Rürup.[14] In his book: "The Philosophers and the Living," the Cologne molecular biologist Benno Müller-Hill reports in the last chapter (entitled "From the Mythology of Animals and Blood to the Cult of Extermination in Auschwitz"), on the "intellectual preparation, the intellectual assistance in the execution and finally the erasing of the traces of the greatest crime ever committed in Germany: the construction of Auschwitz as a place of extermination and production."[15] In lectures, Müller-Hill described to his students how race researchers and biologists assumed from the outset that there were inferior elements in society and that Germany was doomed to extinction by the decline of its race (allegedly organized by Jews). The 1921 volume "Menschliche Erblichkeitslehre und Rassenhygiene" (Human heredity and racial hygiene) by Bauer, Fischer and Lenz became the textbook that shaped public opinion for the next two decades.

---

[13] Julia Amalia Heyer, Entfesselter Hass, in: Der Spiegel, No. 9/23.2. 2019, p. 92.

[14] Dunkle Nazi-Zeit unvergessen. Max-Planck-Gesellschaft wird 100 (Dark Nazi era unforgotten. Max Planck Society turns 100). https://www.n-tv.de/wissen/Max-Planck-Gesellschaft-wird-100-article2330381.html.

[15] Benno Müller-Hill, Die Philosophie und das Lebendige. Campus Verlag, 1981, pp. 199–212. (translated)

One asks oneself again and again how such cruelties and inhumanities, such as the segregation, eradication of Jews, Sinti and Roma and mentally ill people, could have happened. There was a tendency in many cases to blame Hitler alone for those great crimes. Certainly, Hitler was a particularly cruel and brutal man. No smear test can or should be taken on his guilt. But he had too many willing helpers. And it is precisely when we look at various scientists that we clearly see the involvement of pioneers in science. In 1920, the psychiatrist Prof. Dr. Alfred E. Hoche and doctor of law and philosophy Prof. Binding published an article titled: "Die Freigabe der Vernichtung lebensunwerten Lebens" (The release of the destruction of life unworthy of life). They become thereby the crucial pioneers of the organized mass destruction during the period of German fascism. Thus, the pseudo-scientific basis for medical killing was laid, for a development which was to reach its terrible climax 20 years later, for which psychiatry offered both theoretical and practical space, since everything concerning the desired 'racial purity' and the eradication of 'inferior' hereditary material was to be classified under psychohygiene.

If one does not also look at the scientists, then an essential piece is missing when seeking to answer the perennial question: "How could this happen?" What is then also missing is an essential piece of prevention, to ensure that it never happens again. Of course, it is not science in itself, but rather the social structures in and for which science is effective. On 14 July 1933 the Nazi regime enacted the "Law for the Prevention of Hereditary Diseases of Young Persons." It allowed forced sterilization for "congenital feeble-mindedness, schizophrenia, circular (manic-depressive) insanity, hereditary falling sickness, hereditary St. Vitus' dance, hereditary blindness, hereditary deafness, severe physical deformity, and severe alcoholism. As Müller-Hill points out, experts had already prepared a similar bill during the Weimar Republic. Some psychiatrists, such as Prof. Ewald and also the Catholic Church openly spoke out against this law. However, such objections, including proclamations from the pulpit, remained ineffective.

I got to know Benno Müller-Hill personally, when I was asked by my friend Samuel Mitja Rapoport to discuss amongst us three the manuscript and theses of Müller-Hill's book, "The Philosophers and the Living." In 1981, I wrote to Benno Müller-Hill about his resolute confrontation with racism: "I would like to say here once again that what impressed me most was the presentation of the continuous development of racism in exploitative societies—the continuity from Plato's theory of the state to Häckel's failures, to the fascist ideology with the consequence of Auschwitz."[16]

Since Ernst Häckel was introduced to me repeatedly—during my high school graduation, but also during my studies of biology, as well as in the Häckelhaus in Jena—as being a famous biologist, I was especially affected by the mentioned racist failures of Häckel. In the First World War he had indeed written: "A single finely educated German warrior, as they are now falling en masse, has a higher intellectual

---

[16] Letter by Klaus Fuchs-Kittowski, 1981, in: Correspondence of Benno Müller-Hill, Vol I: 1967–1985, pp. 102–103.

and moral life value than hundreds of the raw natural men that England and France, Russia and Italy confront them with."[17]

Another thought in this book affected me very much. Müller-Hill writes: "Even false ideologies have power. Theories of human beings are therefore not, as many would like to see it, timeless, without history and without consequences. When some have lectured long enough on the degeneration of the breed, others collect gold teeth from the murdered for the state."[18] He also wrote: "Deadly Science - The Elimination of Jews, Gypsies and the Mentally Ill 1933–1945."[19] To discuss this book, I met with S.M. Rapoport and Müller-Hill at the Charité, Berlin's pioneering research hospital. The terrible things that B. Müller-Hill reported at our meeting are addressed in the theses above. However, he also noted that the files on the crimes committed by psychiatrists disappeared at some point in time—simultaneously—in both the post-war Federal Republic of Germany (FRG) and German Democratic Republic (GDR). On hearing this, Rapoport jumped up and shouted in the lecture hall, "these files are made of solid material, they can't just disappear, someone must have arranged and executed it."

Religious Traditions as a Cause of Anti-Semitism?

The Protestant theologian Emil Fuchs wrote a resolute article against anti-Semitism as early as 1920.[20] He continued his commitment to the Jews even during the period of fascism, in his interpretations of the New Testament, which were sent as illegal writings to his Quaker friends and to representatives of the forbidden covenant of religious socialists.[21] He turns against the misinterpretation of Paul, which contrasted Christianity and Judaism and was thus appropriated by the fascists to prepare the ground for the Holocaust.

Saint Paul, in his Epistle to the Galatians (chapter II, verses 15–16) says: "We who are Jews by birth and not sinful Gentiles know that a person is not justified by the works of the law, but by faith in Jesus Christ. So we, too, have put our faith in Christ Jesus that we may be justified by faith in Christ and not by the works of the law, because by the works of the law no one will be justified." With the false

---

[17] Ernst Häckel, Ewigkeit. Weltkriegsgedanken über Leben und Tod, Religion und Entwicklungslehre, Reimer/Berlin, 1915. p. 36 (translated)

[18] Benno Müller-Hill, Die Philosophie und das Lebendige. Campus Verlag, 1981, p. 12 (translated).

[19] Benno Müller-Hill, Tödliche Wissenschaft. Die Aussonderung von Juden, Zigeunern und Geisteskranken 1933–1945, Rowohlt, 1984.

[20] Emil Fuchs, Antisemitismus, magazine: Deutsche Politik, 1920.

[21] Klaus Fuchs-Kittowski, Emil Fuchs – Christ, Sozialist und Antifaschist. Freund des arbeitenden Volkes, in: Beiträge zur Geschichte der Arbeiterbewegung, 4/2016, S. 67 - 165.

identification of Paul's critique of the law with the Torah, the "antithesis of Christianity versus Judaism" was anchored at the center of the doctrine of justification.[22]

Emil Fuchs then writes about the dangerous interpretation discussed here: "But if one repeatedly derives from this a justification for the contempt for the Jews, then one should also remember the many who, along with the apostles from the Jews, were bearers of the movement of Jesus Christ, and there were many, which proves the importance of the Palestinian congregations up to the time of the revolution against the Romans in 69/70 AD."[23]

In the preface to the interpretation: "The Good News according to Luke" by Emil Fuchs, Claus Bernet points out that here Fuchs also rejects the accusation against the Jews. He asks: "Which German theologian in his publications of 1939/40 took such a clear position for the Jews (with the exception of Helmut Gollwitzer and his 'Introduction to the Gospel of Luke', who possibly had taken the exegeses of Fuchs himself or knew about them)?[24]

For Fuchs, the core conflict of the Epistle to the Galatians, which revolves around the solidary relationship between Jews and non-Jews, becomes a foil for criticism of the Aryan racial mania, the German-Christian variety of which culminated, for example, in the efforts to "de-Jewify" the New Testament in the vicinity of the Eisenach Institute for "Research into Jewish influence on German church life." With astonishing clairvoyance, Fuchs develops a new paradigm of Paul's interpretation in antithesis to both the prevailing anti-Semitism and the state conservatism of the theology established in church and university, as the theologian Brigitte Kahl points out.[25]

---

[22] Brigitte Kahl, Paulus und das Gesetz im Galaterbrief - Römischer Nomos oder Jüdische Tora? in: Ulrich Duchow, Carsten Jochum-Bortfeld (eds.): Befreiung zur Gerechtigkeit: Die Reformation radikalisieren [Liberation towards Justice: Radicalizing Reformation (Volume 1), LIT-Verlag 2015, p. 86.

[23] Emil Fuchs: Der erste Brief des Paulus an die Thessalonicher, Galaterbrief und Korintherbrief – Eine Auslegung des Evangeliums im Kontext von Verfolgung und Widerstand (1944-1945), Claus Bernet, Klaus Fuchs-Kittowski (Hrsg.): Verlag Dr. Kovač, Hamburg, S. 158, translated.

[24] Claus Bernet: Vorwort (preface). In: Claus Bernet/Klaus Fuchs-Kittowski (Hrsg.): Emil Fuchs: Die Frohe Botschaft nach Lukas. Eine Auslegung des Evangeliums im Kontext von Verfolgung und Widerstand (1939-1941), Verlag Dr. Kovač, S. 16, translated.

[25] Brigitte Kahl, Emil Fuchs` Römerbriefauslegung im Kontext gegenwärtiger Pauluskontroversen, in Gerhard Banse, Brigitte Kahl, Jan Rehmann (Hrsg.): Marxismus und Theologie. Materialien der Jahrestagung 2018 der Leibniz-Sozietät der Wissenschaften, Abhandlungen der Leibniz-Sozietät der Wissenschaften, Band 55, trafo Wissenschaftsverlag, Berlin, 20 – 19, S. 71–80.

Causes of Anti-Humanism, Racism, Anti-Semitism, and Neo-Nazism in the Contemporary World of Work; A Deeper "Perception" of Life and Human Beings is Also Necessary in the Economy!

We are currently experiencing a strengthening of Nazi, i.e., neo-fascist, thinking in society, especially in companies and following similar developments within scientific institutions. In her book: "The Snow of Yesterday is the Deluge of Tomorrow,"[26] Daniela Dahn clearly illustrates the danger of the present situation and the reasons for this Right turn, including within scientific institutions. Under the heading: "Universities" Dahn writes: "Before right-wing extremism reached the center of society, it came from the center of the state. A particularly serious example of this was the award of the first honorary doctorate after the fall of the Berlin Wall by the Humboldt-Universität to the General Staff Officer and Commander of the "Goetz von Berlichingen" SS Panzergrenadier Division [Wilhelm Krelle]. This happened despite the protests of students and others." I too protested vigorously, and wrote a protest letter on behalf of former professors of the Humboldt-Universität, but without success. The powers of the past must have been strong again, even in scientific institutions, to grant (notwithstanding his work in economics, mathematics, and physics) an honorary doctorate to such a divisive figure as Krelle.

That right-wing extremism has now reached the center of society obviously provides a breeding ground in the current world of work. In a large number of publications on the subject, there is talk of the successes of right-wing populism, which were achieved "in part above average" even among trade union members.[27] In practice, today's political language speaks of "right-wing populism"—probably to avoid hastily labeling those who are sympathetic (or potentially susceptible) to far-right ideology, and in the hope of ultimately winning them back into moderate politics. From that perspective, such a strategy may be regarded as appropriate (or necessary), but for me it is a trivialization. What we are confronted with today throughout Europe and in the USA, as a clear shift to the right and the anti-humanist thinking associated with it, follows the pattern already successfully practiced by the German fascists: real grievances are taken up and openly criticized, but without having a solution to deal with them. Therefore, a guilty party is then sought. In those days it was the Jews; today it is foreigners, but also—step by step, and not only in Germany—also again the Jews. The leaders, at least—the ideologists of the movement—are therefore demagogues and not populists.

During Hitler's rise to power there were an extraordinarily high number of unemployed people in Germany. This was an essential basis for anti-Semitism to gain a

---

[26] Daniela Dahn, Der Schnee von Gestern ist die Sintflut von Morgen – Die Einheit– Eine Abrechnung, Rowohlt Taschenbuch Verlag, Hamburg, 2019.

[27] Dieter Sauer, Ursula Stöger, Joachim Bischoff, Reinhard Detja, Bernhard Müller, Rechtspopulismus und Gewerkschaften - Eine arbeitsweltliche Spurensuche, VSA: Verlag Hamburg, 2018, blurb.

foothold among the working class. Today, at least in Germany, there are far fewer unemployed people. Therefore, one first feels a general astonishment about the high degree of neo-fascist thinking, specific expressions and activities in today's working world. There—as it were, under the cover of a much-praised economy of success— the situation has become more acute. This results in losses of control and perspective. Political solutions are rare, and growing criticism of the establishment is the result. Having evaluated a whole series of sociological studies conducted within German companies, Dieter Sauer concludes that "the de-diabolization of the extreme right is progressing."[28]

The defense against this development must be supported by a civil society that has been shaken awake, and must be must be opposed by a labor policy that counteracts the deterioration of conditions in the world of work. This is therefore a task for society as a whole. But science also has a central responsibility here.

Workers have the right to scientifically substantiated statements about real conditions in the world of work. They have a right to see that the world of work is made humane. This is particularly aimed at computer scientists, ergonomists and organizational developers. A human-oriented introduction of modern information and communication technologies requires a socio-technical design of the working world, an information system, work and organizational design from a holistic perspective. This is a major scientific challenge, which is not easy to meet either theoretically, methodologically, or practically, and which requires a great deal of effort, starting with training in the scientific fields directly involved in shaping the world of work.

A really deep, deepened, or new "perception" of the living and the human being is also urgently needed in the economy, because it is precisely this perception which, under the pressure of globalization and digitalization, is even more strongly forced to innovate in order to bring new products and services onto the international market. This calls for ever more research and new knowledge. Thus, the usability of knowledge must be questioned. But this does not have to be connected with the degradation and objectification of man and all living things, as for example through the identification of automaton and man, which is almost unnoticed by many and therefore widespread. This, among other things, should be combined with a revival of the discussion about the possibility of super- or full automation in connection with the development of industry 4.0. The associated degradation of man—not only to the level of an animal as a result of racism, but beyond, i.e., to that of a machine, can have disastrous consequences. Automatic machines (robots) can completely take over human activities and entire production sections, e.g., in the automotive industry. Today, however, hardly anyone considers the complete elimination of humans from production processes (a factory without humans) to be really desirable and possible. The new possibilities of automation, via the Internet of Things (Cyberphysical Systems) and with the support of learning robots, demand and enable a meaningful combination of automaton and human being.

---

[28] Ibid.

The fear of job losses due to structural changes in industry and the service sector is real, e.g., with the planned switch from diesel and petrol engines to electric motors and other mobility concepts. VW's management clearly states that less manpower is needed to build electric vehicles. At the same time, it is announced that VW is in contact with Amazon to develop and introduce a concept for the electronic control of production processes in all plants, which again would result in the loss of large numbers of jobs, this time mainly in administration. The implementation and announcement of permanent restructuring measures in companies are indeed a real source of uncertainty and fear. This is of concern not only in the automotive industry, but also in the expanding logistics and telecommunications sector as well as banking institutions, where staff reductions result in increased pressure to perform.

These structural changes are usually associated with the use of modern information and communication technologies (ICT). This use, however, irrespective of the fact that these general structural changes are catalyzed by the use of ICT, also has further-reaching effects on employment relationships, qualification requirements, and new and increased performance controls in the context of digitization and the use of modern information and communication technologies.

It is part of the responsibility of science to ensure that the use of these methods and technologies is not only technology-oriented, but primarily people-oriented.[29]

People Must Be Able to Recognize the Meaning and Purpose of Their Existence

Wherever the reasons may lie for abandoning the topic of "humanizing work" or work design within the framework of computer science and also within the sub-discipline of computer science and society, it is in any case in keeping with the spirit of the times. It is an example of how even mainstream thinking can succeed in establishing itself at universities. "Even the scientific mainstream sometimes follows the opinion of the powerful."[30] The philosopher Axel Honeth describes very vividly where this farewell or this turning away of science or scientists, including computer scientists, has led to. He writes in an article in the "Deutsche Zeitschrift für Philosophie" on the development so far: "Never in the last two hundred years have efforts to defend an emancipatory, humane concept of work been as badly off as they are today. The actual development in the organization of industrial and service work seems to have undermined all attempts to improve the quality of work..."

Honeth continues: "what is happening in the de facto organization of work, the tendency towards the return of socially unprotected temporary, part-time and

---

[29] Klaus Fuchs-Kittowski, Information, Organisation and Information Technology - Steps towards the development of a human-oriented methodology of information system, work and organisation design. In: Wolfgang Coy, Peter Schirmbacher (eds.): Computer Science in the GDR - Conference Berlin 2010, Humboldt University Berlin, http: edoc.hu-berlin.de/conferences/iddr2010/ pp. 7–36.
[30] Christian Felber, Gemeinwohlökonomie, Piper, 2018, S. 135, translated.

home-work, is also reflected in a twisted way in the shift of intellectual attention and social interests: disappointed were those who forty years ago still placed all their hopes in the humanization or emancipation of work, turning their backs on the world of work in order to turn their attention to completely different, non-productive topics.[31]

A crazy society is essentially the cause of an irrational ideology. If we know today that with the further rigorous exploitation of nature and especially with the further nuclear armament and the development of autonomous weapons; that the existence of mankind as a whole is at stake, but that no decisive counter-movement with a credible and effective alternative is developing; but that society is so crazy that further overexploitation of nature is being carried out and especially a new armament is being started, then one should not be surprised if politically short-sighted people, in view of such a crazy society, also resort to irrational ideologies.

Only where people have a goal, where there is a perspective for the development of society, are their creative powers awakened. Can they, contrary to the widely prevailing egoism and greed, bring about new community in the family, in the world of work and in society, new forms of spiritual life, new knowledge of truth, and thus produce a science and art, forms of working life that guarantee greater justice, social security and peace?

Peace Must Be Secured!

Most people want peace. A life in peace is the first human right! They are therefore in agreement on the goal to be achieved. The differences in concrete wanting do not refer to the goal of wanting, but to the way by which the goal is to be achieved. Peace is to be maintained through armament and deterrence or through negotiations and alliance policy—which, the realization of the situation tells us, is the best way to maintain peace.

If an imperialism of instrumental reason rules (Max Horkheimer, Joseph Weizenbaum), the dominance of a technical–rational reason, which is connected with social rule, then great errors can be produced. For one relies on mathematical calculations where it is necessary to make a proper assessment of the social situation. Particularly important factors, especially human factors, are not or cannot be included in the model calculation. "Extremely important words are missing in the everyday vocabulary of modernity. There is a crucial lack of critical thoughts, ideas that have to do with people, with life in the current practice of everyday affairs in our world," warns Weizenbaum.[32]

---

[31] Axel Honneth, Arbeit und Anerkennung – Versuch einer Neubestimmung. In: Deutsche Zeitschrift für Philosophie (Berlin). (2008)3, S. 327–341, translated.

[32] Joseph Weizenbaum, Lecture at the Occasion of the Dagmar and Václav Havel Foundation VIZE 97 Prize, October 5, 2002: WIDER DEN ZEITGEIST!

The termination in 2019 of the INF (Intermediate-Range Nuclear Forces) Treaty between the USA and Russia shows how quickly military confrontation can escalate again. The treaty made it possible at the end of the 1980s to disarm and ban nuclear short- and medium-range missiles. Now, the political situation has worsened again. Even in Germany, the development of its own nuclear weapons is being discussed. There is already talk of a new Cold War! After the atomic stalemate and the world-wide efforts to create a world free of nuclear weapons—initiatives based on respon-sibility and knowledge—hardly anyone expected a renaissance of nuclear weapons. But the issue will obviously determine security policy in the coming years. If the hardliners should prevail, the end of the relatively peaceful period is imminent.

It is a crucial responsibility of science and scientists to oppose such a develop-ment and to mobilize people against it. There must never be another world war! A war in which the terrible weapons of nuclear fission and fusion, based on missile technology or information and communication technologies, are used. From this arises the responsibility and extraordinary challenge for contemporary science, especially physicists, to establish an effective control system for the abolition of weapons (see roll call from Berlin).[33]

However, the newspapers are reporting these days that the arms race is already underway. A new missile defense system using hypersonic weapons has been announced. "For both medium-range missiles and hypersonic missiles, the advance warning time is so short that a serious clarification of the situation no longer seems possible from a military point of view. There is simply no time to determine whether an attack has begun—or whether there is perhaps only a glitch on the opposing side. A war by accident is one of the horror visions of all military personnel," write Marina Kormbaki and Stefan Koch in the Berliner Zeitung from Washington.[34] The situation has obviously become much worse than that in which the computer scien-tist David Parnas, out of personal responsibility for the "Strategic Defense Initiative" initiated by Ronald Reagan during a tense phase of the Cold War, refused to take part; in which the computer scientists Klaus Brunnstein, Wilhelm Steinmüller, Klaus Haefner and others appealed to Germany's Federal Constitutional Court, stat-ing that in such a situation the Federal President can no longer ensure the protection of the population. Therefore, we should not forget this tradition, but continue it. What can we, what should we do? We need a time of peace and reason, in which the problems of our world are not confronted simply by resort to war and violence, but in accordance with international law, through negotiation.

---

[33] Appell aus Berlin – Für ein kontrollierbares Abkommen zur Abschaffung aller Atomwaffen (Appeal from Berlin - For a controllable agreement to abolish all nuclear weapons). In: Günter Flach, Klaus Fuchs-Kittowski (Hrsg.): Vom atomaren Patt zu einer von Atomwaffen freien Welt – Zum Gedenken an Klaus Fuchs, Abhandlungen der Leibniz-Sozietät, trafo wissenschaftsverlag, Berlin 2012, S. 483–484.

[34] Marina Kormbaki and Stefan Koch in the Berliner Zeitung No. 17, 21 January 2019, p. 2, translated.

To achieve such a time of peace, as Albert Schweitzer[35] emphasized in his Nobel Peace Prize acceptance speech, a peaceful attitude is needed, otherwise the international organizations cannot be effective in peacemaking. Any form of anti-human-ism—the degradation of man to animal or machine—is a blow to such an attitude.

Science Can Follow Its Humanistic Obligation to Serve Life, to Serve Mankind

A possible crisis of science can, as J. Mittelstraß has pointed out, arise from the threatening distance between the production and use of knowledge, which damages the knowledge process by making knowledge lose its very essence, "namely to be an expression of the epistemic essence of man."[36] If we speak of the responsibility of science for guaranteeing human rights, in the fight against the degradation of the living, racism and anti-Semitism; and have shown that science itself—by failing to draw the right conclusions, by stopping at the methodologically necessary reduction or by giving in to excessive compulsion to exploit, with the "threatening distance between the production and use of knowledge"—can at least become a catalyst for inhuman ideologies that have developed in society, then it must be noted that science wants to gain methodologically secured knowledge. "The consoling thing is that this hope is on good ground, the ground of a powerful, incredibly inventive scientific mind, which is able to resist even its confused interpreters and false friends, and of a reason which is still judgmental when it only opposes its own inclinations instead of loving its strengths and weaknesses.[37]

Of all the possible forms of human knowledge, it is science that fundamentally endeavors to gain methodologically sound knowledge. It is therefore right to defend itself against non-scientific influence. It therefore also objects to research projects being hindered by value judgements that are not rationally justified.[38] This cannot mean, however, that a positivist position is taken and that every ethical evaluation of scientific activity is rejected. As Manfred Eigen and Ruth Winkler have said, social information is not protected by "an automatic lock against misuse for self-destruction of life."[39] From this they draw the decisive conclusion that: "an ethics must be oriented towards the needs of humanity. It must guarantee the preservation of

---

[35] Albert Schweitzer, Das Problem des Friedens in der heutigen Zeit, Speech at the reception of the Nobel Peace Prize in Oslo, on 4 November 1954, Verlag C.H. Beck.

[36] Jürgen Mittelstraß, Krise des Wissens? – Über die Erosionen des Wissens- und Forschungsbegriffs, Wissen als Ware, Information statt Wissen und drohende Forschungs- und Wissenschaftsverbote. In: Sitzungsberichte der Leibniz-Sozietät, Band 47, Heft 4, 2001, pp. 21–42.

[37] Ibid p. 40.

[38] Ibid pp. 289–290.

[39] Manfred Eigen, Ruth Winkler, Das Spiel - Naturgesetze steuern den Zufall, R. Pieper & Co Verlag, Munich, 1975, p. 289.

humanity without unduly restricting the individual freedom of each person. Such ethics cannot be derived from any laws of matter below the level of human organization."[40]

Scientists like to elevate "objectivity" to the sole, highest value of science and to reject any further evaluation. This makes everything seem feasible. But from the shattering experience that this has paved the way to a deadly science, we must learn that it is not permissible to do everything that can be done. The urgent warning to be vigilant, the inescapable invitation to repeatedly examine what can truly be considered truth, scientific knowledge, arises for us today. Consequently—contrary to what is often postulated—"pure objective knowledge" is not the sole highest value of science. The search for true knowledge—real insight—also encompasses the question of the criterion of truth. To reconsider the criterion of truth is scientifically founded practice. However, this means that "pure objective knowledge" ceases to be the sole and highest value of science; in addition, there are rationally based ethical values. Thus, science can follow its humanistic obligation to serve life, to serve humanity.

---

[40] Ibid.

Chapter 11
The Ambivalences of the Digital—Humans and Technology Between New Dreams/Spaces of Possibility and (Un)Noticeable Losses

Hartmut Graßl, Stefan Bauberger, Johann Behrens, Paula Bleckmann, Rainer Engels, Eberhard Göpel, Dieter Korczak, Ralf Lankau, and Frank Schmiedchen

> *The need for a major public debate on the ambivalences of the digital is actually coming almost too late, because the misuse of large amounts of data has already begun.*
>
> Hartmut Graßl

Abstract Every new technology is used by us humans almost without hesitation. Usually the military use comes first. Examples from recent history are the use of chemical weapons by Germany in the First World War and of atomic bombs in the Second World War by the US. Now, with the rapid advances in microelectronics over the past few decades, a wave of its application, called digitization, is spreading around the world with barely any control mechanisms. In many areas this has simplified and enriched our lives, but it has also encouraged abuse. The adaptation of legislation to contain the obvious excesses of "digitization" such as hate mail and anonymous threats is lagging behind massively. We hear almost nothing about technology assessment through systematic research; it is demanded at most by a few, usually small groups in civil society, which draw attention to the threats to humankind—future and present—and the Earth's ecosystem. One such group, the Federation of German Scientists (VDW) e.V., in the spirit of the responsibility of science for the peaceful and considered application of the possibilities it creates, asked three of its study groups to jointly organize its 2019 Annual Conference. The study groups "Health in Social Change," "Education and Digitization," and

H. Graßl (✉) · S. Bauberger · J. Behrens · P. Bleckmann · R. Engels
E. Göpel · D. Korczak · R. Lankau · F. Schmiedchen
The Federation of German Scientists (VDW), Berlin, Germany
e-mail: info@vdw-ev.de

"Technology Assessment of Digitization" formulated the following position paper for the 2019 VDW Annual Conference, entitled "Ambivalences of the Digital."

Introduction[1]

The Federation of German Scientists (VDW) e.V. perceives the process of accelerated digitization and interconnectedness as well as the development of so-called "artificial intelligence" (AI, also: machine learning) as the potential for epochal social change. As with the development of the use of tools and the spread of language, writing, and printing, a fundamental change in the organization of social life can already be observed today as a result of digitization. These changes hold both opportunities and threats for humankind—future and present—and the Earth's ecosystem. The VDW therefore recognizes the necessity and urgency of critically examining digitization as a series of technological processes and their social prerequisites, applications, limits, and consequences.

It is part of the VDW's identity to critically examine new technologies with regard to hazard potential, and to initiate risk analyses, in addition to proposing measures to avert hazards at an early stage.

In light of the excitement, almost religious in nature, regarding digitization, interconnectedness, and AI in practically all areas of life, the VDW considers it its task to point out underestimated or ignored scientifically and socially highly relevant existential problems associated with this development and to make well-founded proposals for ethically acceptable means of handling them.

For example, the VDW considers one of the immediate dangers to be the creation of completely new, long- lasting, unpredictable, profound dependencies for individuals, institutions, and states—that only few can escape—of digitization, interconnectedness, and AI. This development has the potential to deepen existing social inequalities in societies and to intensify the global discrimination of particularly low-income population groups. As such, digitization, interconnectedness, and AI pose a sustainability risk. Whether they prove to be socially and ecologically harmful depends primarily on the results of social, political, and economic negotiation processes, struggles, and decisions.

In addition to these societal challenges, there are also technology-related risks that need to be addressed comprehensively and rapidly, regardless of the societal embeddedness of technology use. These include, among others, objective limits to the quantity of use and security issues in AI development, but also in particular the hidden manipulation of users through processing and exploitation of unmanageable

---

[1] The text corresponds to the position paper of the same name, which was prepared for the VDW Annual Conference 2019. The VDW study groups "Health in Societal Change," "Education and Digitisation," and "Technology Assessment of Digitisation" participated in preparing this statement. The ten theses presented at the end formed the basis for discussion at the annual conference.

amounts of data. These are used by corporations to control consumption, by political groups for disinformation, and by authoritarian states for oppression and social control.

Correlations within mass data, discovered with the help of AI, are also confused with causalities and such correlations are being used to judge people in various ways and as the basis for subsequent decision making. The available experience shows that this repeatedly leads to social disadvantages and discrimination (structurally, too) and that these misjudgments (e.g., when looking for a partner, a job, housing, or a loan) cannot be corrected or can only be corrected with great effort. Such processes can accelerate downward spirals of social exclusion.

The VDW observes with great concern that the digitization of more areas of human life questions the self-image of individuals and societies such that fundamental threats to the health, dignity, and freedom of a large part of humanity are looming and democratic societies are endangered.

Underlying Human Image

The current development of digital technologies and machine learning is based on a reductionist world view, which comes to a head in the metaphor of people as information processing systems. This image already has its origins in cybernetics and behaviorism, which understand humans merely as an organic feedback system. According to this, any behavior is only the more versus less appropriate result of a neuronal evaluation of sensory data. Characteristics such as consciousness, freedom, or the self are at best regarded as phenomena emerging from information or from the activity of the neural substrate. The question of the meaning of an individual's being as a constitutive moment of human existence remains completely disregarded in this perspective.

Human beings thus reduced to their mechanical characteristics become deficient beings whose cognitive abilities, although considered exceptional in the animal kingdom, are ultimately inefficient due to their organic limitations and should therefore be optimized by adequate technological supplementation. For many experts involved in the development of digital technologies, AI is not just a simple extension of human cognitive competence, but rather a new branch of the evolution of intelligence itself, in which humans are finally overtaken and overcome by AI, which is becoming independent of them.

In the last consequence, human intelligence is reduced to propositional thinking, which is also postulated by representatives of this view as the only reliable basis for all decisions. Whether in road traffic or in the choice of a life partner, decisions are nothing more than neurally mediated environmental analyses that ultimately lead to true or false results. However, since such algorithmically comprehensible calculations can in principle be carried out much more efficiently and accurately by "intelligent" machines, it is not a contradiction, but rather a compelling consequence to leave decisions with far-reaching social consequences to AI in the future.

If an ultimately inhuman and misanthropic worldview remains the dominant view in the future, which guides the development and application of digital technologies and the closely associated image of AI, then digitization will become an immanent threat for humankind and possibly for organic life on Earth per se. From the VDW's perspective, such a reduced view of humankind does not correspond to the actual nature of humankind, nor to its dignity and intrinsic value. Through their existence, humans are already the answer to the question of meaning, which they pose as conscious beings. It is not the human who must adapt to technology, but rather technology that must always remain a tool of humans, to be used for their benefit and the benefit of all creatures.

Health and Social Participation

According to the definition of the World Health Organization (WHO),[2] the VDW understands health as a bio-psycho-social construct, including self-determination and participation in the life of society. The interconnectedness of machines capable of learning can contribute to human health, but can also harm it. On the one hand, as tools, interconnected machines can be used for the coordination of care, for instance at the interface between outpatient and inpatient care, for faster diagnostics, better prosthetics, or, by reducing the workload, to creating time for more humane care, therapy, and medicine. On the other hand, initial experience shows that interconnectedness through learning machines exacerbates existing social problems. We observe monopolistic appropriation, manipulation, and advertising paternalism, division, and discrimination, and altogether a departure from the humane, as already indicated above.

Although research and development underlying such technologies are usually publicly funded, the intellectual property in the form of the equipment, algorithms, and source codes are declared private property and trade secrets. Likewise, users' private data are used and misused by private companies and government agencies in at least immoral and often illegal ways.

In the combination of (depth) psychological findings and the use of digitalized health applications, people are seduced into unreflected behavior controlled by automats, which can cause them harm. In particular, the radical-utilitarian approach to human self-optimization, including the optimization of one's children (already prenatally), as a suggested prerequisite for a successful and happy life, can cause illness and social damage.

---

[2] "Health is a state of complete physical, mental and social well-being and not merely the absence of disease or infirmity." http://apps.who.int/gb/bd/PDF/ bd47/EN/constitution-en.pdf?ua=1 (30.10.2019). The definition is set out in the World Health Organization (WHO) declaration. It was adopted by the International Health Conference held in New York from 19 June to 22 July 1946 and signed on 22 July 1946 by the representatives of 61 states. It came into force on 7 April 1948.

Disease and death are not only biological–medical processes, but always integrated into social processes. Humans must therefore always be viewed holistically and not in a quantitative–reductionist way.

The VDW considers dangers in the fact that the tendency to turn away from the humane also has serious consequences regarding human health. If humans are increasingly adapted to the technical requirements of machines, as demonstrated for about 100 years exemplified by assembly line production, and this pressure to adapt becomes increasingly all-encompassing, then well-known tendencies to perceive "lower-functioning people" as either reparable (e.g., through rehabilitation measures), or to segregate them as avoidable defects (e.g., in the case of genetic defects) will increase.

Education and Digitization

The educational policy and educational science discourse on teaching, learning, and digitization is largely dominated by media education, IT (learning media) development, and quantitatively oriented empirical educational research. The focus is one-sidedly on the opportunities offered by the digitization of teaching media and the personalization that this makes possible, including small-scale learning assessment for learning control. The purported opportunities offered by this form of mediatized schooling are often claimed but are only partially proven or verifiable. On the other hand, by largely excluding the perspectives of other fields of research, the current discourse fails to sufficiently consider risks. Examples include the important and critical contributions from historical and philosophical educational research, media addiction research, pediatric and developmental psychological research into the effects of media, public health and prevention science, neurobiology, attachment research, the criticism of algorithm-based control systems and data exploitation economics, as well as research on the effects of non-ionizing electromagnetic radiation.

A no less problematic but less visible constriction of discourse occurs if the proposed solutions to "digital risks" do not take place on a broad level oriented on the design of a humane environment. It is highly problematic that until recently the politically discussed approaches were limited to purely technical improvements (e.g., better encryption for more protection of student data) or to the level of self-optimization of individuals in terms of teaching "digital risk avoidance skills." This is in clear contradiction to the findings of prevention research, according to which relational prevention, i.e., the creation of healthy living spaces, can contribute more to the prevention of risk behavior than behavioral prevention, which starts with the behavior of individuals. This is even more relevant where younger target groups are concerned.

The VDW therefore sees an urgent need to finance robust, transdisciplinary, and independent technology assessments (TA) to compare different technology paths in educational institutions and to ask whether children should be introduced to technology as early as possible ("Early High Tech") or whether, in accordance with their physical and cognitive development, real (sensorimotor) life experiences should be the initial focus of attention ("Early High Touch; High Tech later").

The overarching control of educational processes (educational governance) at country and institution level must remain in the hands of the people and legitimized by democratic decision-making processes, and must not be delegated to Big Data-based systems, as is already largely the case in the US and some other countries.

Based on the present state of knowledge, the VDW sees an urgent need for financing the development and implementation of modern teaching concepts that are oriented on the development phases and lay the foundations for media literacy until the end of kindergarten without any digital screen media, and until the end of primary school largely without any digital screen media (e.g., through an "Analogue Pact#D"). Furthermore, it advocates the financing and implementation of modern, non-commercial concepts for secondary schools and universities for the use of digital media for teaching and learning, as well as the creation of a digital infrastructure for these facilities that cannot become a control and management technology for the users and does not generate long-term learning or personality profiles.

Economy, Labor, Society

The developments referred to as the "fourth" industrial revolution, which are characterized by accelerated digitization and interconnectedness of production, logistics, trade, and services, as well as the increasing use of AI, are once again changing the way we do business, across the entire global value chain.

This results in serious challenges for the economy, society, and state in order to make inclusive use of the opportunities for social and ecological transformation potentially contained therein and to effectively ward off the dangers of deepening social inequalities and spreading unsustainable consumption patterns. The disruptive nature of the changes increases the pressure to comprehensively identify and swiftly implement the necessary political and legal decisions and measures. From the perspective of the VDW, the focus should be on improving the enforcement of economic, social, and cultural human rights (e.g., effective protection of competition, safeguarding and further development of core labor standards, functioning solidarity systems for social security). How can potential opportunities for an improved relationship between gainful employment, social/communal involvement, and time for oneself, family, and friends be used?

Already today, the state of digitization could facilitate enormous freedom for the organization of economic and social systems. This will probably increase in the coming years. At the same time, however, it has also become apparent for some time now how the ongoing digitization in some countries is being used to create new private or state monopoly structures and to reduce civil liberties in favor of centralized state power, the suppression of foreigners and dissenters, and the promotion of economic–behavioral uniformity.

In contrast, the VDW considers an enormous welfare potential in the further advancement of mixed economic systems in which public, cooperative, and private actors (above all small- and medium-sized enterprises) provide different contributions to solving the challenges of rapid technological developments and the changing framework and contributing conditions (e.g., climate change, demographic development). In the VDW's view, a broad diversity of actors and objectives (e.g., the provision of public goods, orientation towards the common good, private-sector profit motive) has in the past proved to be rather crisis-resistant.

This welfare potential can unfold its effect in favor of social cohesion to the extent that it effectively strengthens lower and middle incomes. This results from the different savings and consumption rates of the various income groups. To the extent that productivity gains arise from the increasing digitization, interconnectedness, and use of AI, new ways must be sought to sustainably finance existing or developing social security systems. This cannot be achieved by a one-sided burden on the production factor labor (payroll taxes, etc.). Energy consumption, capital income, and assets must be used to finance this. Tax-based basic funding in combination with citizens' or workers' insurance also makes it easier to switch between the tasks people devote themselves to, such as dependent employment, self-employment, involvement in family, community, and society (e.g., support for children, young people, people in need of care, and the elderly). These systems will only be successful, however, if they also consider cross-border solidarity in order to better safeguard individual wishes and the need for mobility of employees. In the EU, for example, a first sensible step would be to introduce a common European unemployment insurance scheme. Such a fund could help to arm member states against economic crises and the high unemployment that accompanies them.

The changes for gainful employment can be used to address each person's desire to be useful in the economic system and to combine this, better than before, with the need for socially necessary work, and to finance both adequately. The question of which resources we use for which production is in principle a social one, the answer to which can be organized in different ways in the democratic process. For example, the loss of millions of jobs leads to social distortions even if a similar number or even more new jobs are created in quantitative terms, because the workers who have been rationalized away are often not qualified for the new job profiles. Without adequate social measures, which must go beyond primarily vocational training offensives, this can lead to these people feeling marginalized and becoming susceptible to anti-democratic, authoritarian propaganda and pseudo-declarations.

Regulation

In order to counteract any form of abuse of accelerated digitization and interconnectedness as well as the development of AI, clear, binding and enforced ethical and legal objectives, standards, policies, and regulations are necessary which secure and expand democratic (including civil society) control, monitoring, and participation opportunities and rights.

Essential and consequential decisions must be based on the results of technology assessments. This may also require bans to be imposed in areas where the current state of research indicates a negative risk–reward-balance. For example, the available research results demonstrate that the use of digital screen media during kindergarten and primary school age has predominantly negative consequences, so that a moratorium on the use of digital screen media by such young learners is necessary.

The Asilomar Principles for the development and use of artificial intelligence, which have become Californian law, are only a rudimentary approach[3] and must be further developed in a social and democratic process, adapted to reality, tightened, and consistently applied. For example, the transparency of algorithms must be ensured: Algorithms that decide on the life paths and life chances of people (e.g., in school, at work, in health, in justice) must be disclosed and the underlying calculation processes must be comprehensible.

Further development of existing regulations for the effective enforcement of core labor standards in changing industrial relations is just as necessary, as adjustments in competition law, in regulations for the protection of intellectual property rights, and in the improvement of social security systems (e.g., trade union protection in platform economies, taxation of data use, data protection and privacy protection, financial participation of people in the use of their data, immorality of certain data transfer agreements).

Ten Key Questions

1. What image of the world and humankind do we have and promote, and what influence may technological developments have on it?
2. What utopias do we aspire to in the coming decades and what role should digitization, interconnectedness, and AI play in this? For example, do we want progressively more decisions that fundamentally affect human existence to be made by machines?
3. Which (ethical) maxims and limits do we want for digital extensions and alleged "improvements" of humans (prenatal genetic interventions; monitoring and control of vital functions)?
4. How can and should the economy, state, and society guarantee the establishment and expansion of a non-commercial, digital infrastructure?
5. How can social participation of all be ensured?
6. How can the use and added value of digital media in the classroom be objectively determined and democratically evaluated?

---

[3] The VDW Study Group on Technology Assessment of Digitisation has published a statement on the Asilomar principles of artificial intelligence. This is available as a download on the VDW website under the following link: https://vdw-ev.de/wp-content/uploads/2018/05/Stellungnahme-SG-TA-Digitalisierung-der-VDW_April-2018.pdf (1 January 2022, German only).

7. How can elementary cultural techniques (reading, writing, arithmetic, making music, working, drawing, etc.) and the basics of knowledge and skills (logical thinking, language skills, understanding of connections, concentration, attention, etc.) be successfully taught and thus preserved in a verifiable way?
8. How can the fourth industrial revolution be used for social cohesion (also in the Global South) and ecological renewal? What alternative models for the integration of economic and social policy are possible?
9. How can distortions in structural change (e.g., through massive changes in job profiles) be significantly reduced and made socially acceptable, and how can social systems be efficiently developed?
10. How can we effectively combat and sustainably counteract manipulation by private and state actors and interests?

Federation of German Scientists (VDW) e.V.
Since the foundation of the Federation of German Scientists (VDW) e.V. in 1959 by prominent nuclear scientists, among them Carl Friedrich von Weizsäcker, who had previously spoken out publicly as a "Göttingen 18" signatory against nuclear armament of the German Armed Forces, the VDW has felt committed to the tradition of responsible science. At annual conferences, in interdisciplinary study and project groups, scientific publications, and public statements, it takes a stand on questions of scientific orientation, technological developments, and peace and security policy. The role of science itself is also a subject of consideration, both in the genesis and the solution of problems. Around 350 natural scientists, humanities scholars, and social scientists are organized in the VDW, and work together on current and pressing issues in an inter- and transdisciplinary manner. With the results of its work, the VDW addresses the sciences, the interested public, and decision makers at all levels of politics, society, and the economy.

In accordance with its statutes from 1959, the VDW sets itself the following goals:
– to strengthen the sense of responsibility of scientists for the impact of their research on society
– to study the problems arising from the progressive development of science and technology
– to give a public voice to science and its representatives
– to influence decisions in an advisory capacity and to oppose the misuse of scientific results
– to stand up for the freedom of research and promote the free exchange of its results.

Bibliography

Assis-Hassid, S., Heart, T., Reychav, I., & Pliskin, J. S. (2016). Modelling factors affecting patient–doctor–computer communication in primary care. *International Journal of Reliable and Quality E-Healthcare (IJRQEH), 5*(1), 17ff. https://doi.org/10.4018/IJRQEH.2016010101

Bartels, J. (2017). What's all this silence? Computer centered communication in patient doctor computer communication. In J. Bartels (Ed.), *Health professionals education in the age of health information systems, mobile computing and social networks* (pp. 23–34). https://doi.org/10.1016/B978-0-12-805362-1.00002-4

Behrens, J. (2019). *Theorie der Pflege und der Therapie*. Hogrefe.

Behrens, J., & Langer, G. (2016). *Evidence based nursing and caring: Methoden und Ethik der Pflegepraxis und Versorgungsforschung – Vertrauensbildende Entzauberung der "Wissenschaft"*. Hogrefe.

Bittner, J., Dockweiler, C., & Thranberend, T. (2018). Roadmap Digitale Gesundheit Handlungsempfehlungen für eine Digitalisierung im Dienst der Gesundheit. In U. Repschläger, C. Schulte, & N. Osterkampf (Eds.), *BARMER Gesundheitswesen aktuell* (pp. 62–91).

Bleckmann, P., & Lankau, R. (Eds.). (2019). *Digitale Medien und Unterricht*. Beltz.

Bleckmann, P. (2018). Toward media literacy or media addiction? Contours of good governance for healthy childhood in the digital world. In M. Matthes, L. Pulkkinen, B. Heyes, & C. Clouder (Eds.), *Improving the quality of childhood in Europe* (Vol. 7, pp. 103–119). Alliance for Childhood European Network Foundation.

Bleckmann, P., & Mößle, T. (2014). Position zu Problemdimensionen und Präventionsstrategien der Bildschirmnutzung. *Sucht, 60*(4), 235–247.

Budzinski, O., & Schneider, S. (2017). *Smart Fitness: Ökonomische Effekte einer Digitalisierung der Selbstvermessung*. Institut für Volkswirtschaftslehre.

Bündnis für humane Bildung. (2017). *Sieben Forderungen des Bündnisses für humane Bildung*. http://www.aufwach-s-en.de/wp-content/uploads/2017/10/buendnis_forderungen.pdf

Christiaensen, L., & Demery, L. (Eds.). (2017). *Agriculture in Africa – Telling myths from facts*. World Bank Group.

Deutscher Ethikrat. (2017). *Big Data und Gesundheit Datensouveränität als informationelle Freiheitsgestaltung*. Deutscher Ethikrat.

Dockweiler, C., Kupitz, A., & Hornberg, C. (2017). Nutzerorientierung in der telemedizinischen Forschung und Entwicklung: Welche Potenziale besitzen partizipative Verfahren? In F. Duesberg (Ed.), *eHealth 2017 – Informations und Kommunikationstechnologien im Gesundheitswesen* (pp. 118–122). Medical Future Verlag.

Ferber, L. V., & Behrens, J. (1997). *Public Health – Forschung mit Gesundheits- und Sozialdaten: Stand und Perspektiven. Memorandum zur Analyse und Nutzung von Gesundheits- und Sozialdaten*. Sankt Augustin.

Förschler, A. (2018). Das 'Who is who?' der deutschen Bildungs-Digitalisierungsagenda – eine kritische Politiknetzwerk-Analyse. *Pädagogische Korrespondenz, 58*(2), 31–52.

Frankfurt-Dreieck zur Bildung in der digital vernetzten Welt. (2019). Ein interdisziplinäres Modell.

Frey, C. B., & Osborne, M. A. (2013). *The future of employment: How susceptible are jobs to computerisation*. Oxford University.

Göpel, Eberhard (2017). Eine Gesundheits-Wende in der Medizinforschung ist notwendig! https://vdw-ev.de/prof-dr-eberhard-goepel-eine-gesundheits-wende-in-der-medizinforschung-ist-notwendig/

Gövercin, M., Meyer, S., Schellenbach, M., Steinhagen-Thiessen, E., Weiss, B., & Haesner, M. (2016). SmartSenior@home: acceptance of an integrated ambient assisted living system. Results of a clinical field trial in 35 households. *Informatics for Health and Social Care, 41*, 430–447. https://doi.org/10.3109/17538157.2015.1064425

Hallward-Driemeier, M., & Nayyar, G. (2018). *Trouble in the making? The future of manufacturing-led development*. The World Bank Group.

Hartong, S. (2018). Wir brauchen Daten, noch mehr Daten, bessere Daten! Kritische Überlegungen zur Expansionsdynamik des Bildungsmonitorings. *Pädagogische Korrespondenz, Heft, 58*, 15–30.

Hastal, M. R., Dockweiler, C., & Mühlhaus, J. (2017). Achieving end user acceptance: Building blocks for an evidence based user centered framework for health technology development and assessment. In M. Antona & C. Stephanidis (Eds.), *Universal access in human computer interaction: Human and technological environments* (pp. 13–25). Springer.

ILO. (2017). *Inception report for the global commission on the future of work.* Genf.

Kinderkommission des Deutschen Bundestags. (2019): Stellungnahme zum Thema „Kindeswohl und digitalisierte Gesellschaft: Chancen wahrnehmen – Risiken bannen". http://www.bundestag.de/resource/blob/651028/ 0de1b58a7b242fe62c293a19f00cb055/2019-07-10-Stellungnahme-Kindeswohl-und-digitalisierte-Gesellschaft-data.pdf

Kolany-Raiser, B. (2016). Der Verbraucher als Datenlieferant: rechtliche Aspekte von "smarten" Produkten. In: Verbraucherzentrale Nordrhein-Westfalen e.V; Kompetenzzentrum Verbraucherforschung NRW (Hrsg.): Schöne neue Verbraucherwelt? Big Data, Scoring und das Internet der Dinge, Düsseldorf, S. 47–66.

Korczak, D. (2013). Ist der Erfolg von Alkoholpräventionsmaßnahmen mess- und evaluierbar? In: Suchttherapie, Bd. 14, Nr. 03, Thieme: Stuttgart, S. 114 118.

Korczak, D., Steinhauser, G., & Kuczera, C. (2012). *Effektivität der ambulanten und stationären geriatrischen Rehabilitation bei Patienten mit der Nebendiagnose Demenz, Schriftenreihe Health Technology Assessment (HTA)* (Vol. Bd. 122). DIMDI.

Korczak, D. (Ed.). (2007). *Zukunftspotentiale der Nanotechnologien. Erwartungen, Anwendungen, Auswirkungen.* Asanger.

Kucklick, C. (2014). *Die granulare Gesellschaft. Wie das Digitale unsere Wirklichkeit auflöst.* Ullstein.

Lankau, R. (2017). *Kein Mensch lernt digital: Über den sinnvollen Einsatz neuer Medien im Unterricht.* Beltz.

Lembke, G., & Leipner, I. (2015). *Die Lüge der digitalen Bildung. Warum unsere Kinder das Lernen verlernen.* Redline Verlag.

Lenzen, M. (2018). *Künstliche Intelligenz. Was sie kann und was uns erwartet.* Pieper.

Meurer, J., Müller, C., Simone, C., Wagner, I., & Wulf, V. (2018). Designing for Sustainability: Key Issues of ICT Projects for Ageing at Home. *Computer Supported Cooperative Work (CSCW), 27*(3-6), 495–537.

Misselhorn, C. (2018). *Grundfragen der Maschinenethik.* Reclam.

Mößle, T. (2012). *Dick, dumm, abhängig, gewalttätig? Problematische Mediennutzungsmuster und ihre Folgen im Kindesalter. Ergebnisse des Berliner Längsschnitt Medien.* Nomos Verlag.

Mort, M., Roberts, C., Pols, J., Domenech, M., & Moser, I. (2015). Ethical implications of home telecare for older people: a framework derived from a multisited participative study. *Health Expectations, 18*(3), 438–449.

Müller, R., & Bäumer, M. (2018). *Out of office. Wenn Roboter und KI für uns arbeiten.* Stiftung Historische Museen Hamburg.

Müller, C., Hornung, D., Hamm, T., & Wulf, V. (2015). *Measures and Tools for Supporting ICT Appropriation by Elderly and Non Tech Savvy Persons in a Long Term Perspective* (pp. 263–281).

Dierk, T., & Bahrs, O. (2018). Beiträge der Salutogenese zu Forschung, Theorie und Professionsentwicklung im Gesundheitswesen. In J.-G. Monika & P. Kriwy (Eds.), *Handbuch Gesundheitssoziologie* (pp. 1–28). Springer Fachmedien VS.

Münch, R. (2018). *Der bildungsindustrielle Komplex. Schule und Unterricht im Wettbewerbsstaat.* Beltz-Juvena.

Paternoga, D., Rätz, W., & Pietron, D. (2019). *Eine andere Digitalisierung ist möglich. Chancen und Risiken einer vernetzten Gesellschaft.* VSA.

projekt: futur iii – Digitaltechnik zwischen Freiheitsversprechen und Totalüberwachung: http://futur-iii.de

Ramge, T. (2018). *Mensch und Maschine. Wie Künstliche Intelligenz und Roboter unser Leben verändern*. Reclam.

Rümelin, N., & Weidenfeld, N. (2018). *Digitaler Humanismus*. Pieper.

Schmiedchen, F. et al. (2018). Stellungnahme zu den Asilomar-Prinzipien zu künstlicher Intelligenz. VDW e.V., Berlin. https://vdw-ev.de/wp-content/uploads/2018/05/Stellungnahme-SG-TA-Digitalisierung-der-VDW_April-2018.pdf

Schmiedchen, F., Kratzer, K. P., Link, J. S. A., & Stapf-Finé, H. (Eds.). (2022). *The world we want to live in: Compendium of digitalisation, digital networks, and artificial intelligence*. Logos Verlag.

Sobral, D., Rosenbaum, M., & Figueiredo-Braga, M. (2015). Computer use in primary care and patient physician communication. *Patient Education and Counseling, 2015, 98*(12), 1453–1652.

Spiekermann, S. (2019). *Digitale Ethik. Ein Wertesystem für das 21. Jahrhundert*. Droemer.

Spitzer, M. (2015). *Cyberkrank. Wie das digitalisierte Leben unsere Gesundheit ruiniert*. Droemer.

Techniker Krankenkasse. (2016). Beweg Dich Deutschland! TK Bewegungsstudie 2016, Hamburg.

Tretter, F. (2008). *Ökologie der Person. Auf dem Weg zu einem systemischen Menschenbild: Perspektiven einer Systemphilosophie und ökologisch-systemischen Anthropologie*. Pabst Science Publ.

Tegmark, M. (2017). *Leben 3.0. Mensch sein im Zeitalter Künstlicher Intelligenz*. Ullstein.

Unabhängiges Landeszentrum für Datenschutz Schleswig-Holstein, GP Forschungsgruppe. (2014). *Scoring nach der Datenschutz-Novelle 2009 und neue Entwicklungen*. Kiel.

Weizenbaum, J. (1978). *Die Macht der Computer und die Ohnmacht der Vernunft*. Suhrkamp.

Zierer, K. (2018). Die Grammatik des Lernens, in: FAZ, 4.10.2018, S. 7, https://www.faz.net/aktuell/feuilleton/hoch-schule/digitale-schule-die-grammatik-des-lernens-15819548.html

Zuboff, S. (2018). *Das Zeitalter des Überwachungskapitalismus. Frankfurt am Main*. Campus.

Chapter 12
Heritage Requires Citizens' Knowledge: The COST Place-Making Action and Responsible Research

Heike Oevermann, Ayse Erek, Carola Hein, Conor Horan,
Kata Krasznahorkai, Ida Sofie Gøtzsche Lange, Edmond Manahasa,
Marijke Martin, Marluci Menezes, Matej Nikšič, Paulina Polko, Juli Székely,
Simone Tappert, and Pekka Tuominen

Abstract This chapter reflects on responsible science with an eye toward concrete research practice. To this end, we briefly introduce the RRI paradigm (Responsible Research and Innovation) and then highlight seven EU research projects in the context of a transnational COST Action project. This COST Action will investigate how placemaking activities, like public art, civil urban design, and local knowledge production, reshape and reinvent public space, and improve citizens' involvement in urban planning and urban design, especially in the context of heritage sites. The chapter introduces heritage case studies that either contrast, differentiate, and add to existing knowledge and practices in placemaking through specific initiatives, or enable the establishment of common ground within a wider constellation of societal actors and both, as we see, contribute in different ways to responsible research. We analyze how the four criteria of RRI, namely anticipation, reflexivity, inclusion, and responsiveness are considered and implemented, and the extent to which digital tools are supportive. Obviously, coproduction of knowledge is not sufficient when we call for responsible science in the narrow sense, hence the development of common ground also appears necessary.

H. Oevermann (✉)
Bauhaus-Universität Weimar, Weimar, Germany
e-mail: heike.oevermann@uni-weimar.de

A. Erek
Kadir Has University Istanbul, Istanbul, Turkey

C. Hein
TU Delft, Delft, The Netherlands

C. Horan
TU Dublin, Dublin, Ireland

K. Krasznahorkai
Universität Zürich, Zurich, Switzerland

© The Author(s) 2022
H. A. Mieg (ed.), *The Responsibility of Science*, Studies in History and
Philosophy of Science 57, https://doi.org/10.1007/978-3-030-91597-1_12

This chapter reflects on responsible science with an eye toward concrete research practice. To this end, we briefly introduce the RRI paradigm (Responsible Research and Innovation) and then highlight seven EU research projects in the context of a transnational COST Action project. In this project, citizens' knowledge creates visibility of alternative perspectives that mobilizes for engagement and inclusion. Obviously, coproduction of knowledge is not sufficient when we call for responsible science in the narrow sense, and hence the development of common ground also appears necessary.

A New, Responsible Role of Science Defined by the European Research Initiatives

Responsible Research and Innovation

Responsible Research and Innovation (RRI) is a conceptual framework for integrative research policy, primarily at the EU level. The RRI approach emerged from poor experience with large-scale technology implementation and technology-oriented policy making (Macnaghten & Owen, 2011; Sutcliffe, 2011). RRI is integrative insofar as different streams come together: technology-reflective innovation, gender mainstreaming, opening science for social participation, and the ethics of science. Since about 2010, RRI has found its way into the definition of European research.

I. S. G. Lange
Aalborg Universitet, Aalborg, Denmark

E. Manahasa
Epoka University Tirana, Tirana, Albania

M. Martin
University of Groningen, Groningen, The Netherlands

M. Menezes
National Laboratory for Civil Engineering LNEC, Lisbon, Portugal

M. Nikšič
Urban Planning institute of the Republic of Slovenia (UIRS), Ljubljana, Slovenia

P. Polko
WSB University, Warsaw, Poland

J. Székely
Eötvös Loránd University Budapest, Budapest, Hungary

S. Tappert
University of Applied Sciences Northwestern Switzerland, Basel, Switzerland

P. Tuominen
University of Helsinki, Helsinki, Finland

> *Responsible Research and Innovation means that societal actors work together during the whole research and innovation process in order to better align both the process and its outcomes with the values, needs, and expectations of European society. RRI is an ambitious challenge for the creation of a Research and Innovation policy driven by the needs of society and engaging all societal actors via inclusive participatory approaches.* (European Commission, 2012).

Subsequently, a number of seminal papers have been published on RRI. The definition of RRI as an open process became significant (von Schomberg, 2013, p. 63):

> Responsible Research and Innovation is a transparent, interactive process by which societal actors and innovators become mutually responsive to each other with a view to the (ethical) acceptability, sustainability and societal desirability of the innovation process and its marketable products (in order to allow a proper embedding of scientific and technological advances in our society).

RRI constituted the basis for the SwafS (Science with and for Society) subprogram of the EU research framework program Horizon 2020 (2014–2020). SwafS was "instrumental in addressing the European societal challenges tackled by Horizon 2020, building capacities and developing innovative ways of connecting science to society" (European Commission, 2021). SwafS clearly linked societal issues to innovation: "It makes science more attractive (notably to young people), raises the appetite of society for innovation, and opens up further research and innovation activities." (loc. cit.) SwafS included a monitoring project for RRI (Monitoring the Evolution and Benefits of Responsible Research and Innovation. See: http://morri-project.eu/).

How can RRI be implemented or captured? At the level of EU research, a criteria-oriented approach has become established—for the sake of manageability. The following six criteria or "keys" are considered indicative of RRI (op. cit.), namely: [public] engagement; gender equality; science education; open access; ethics; and governance. Better-founded support for implementation can be found in the "framework for responsible innovation" developed by Stilgoe et al. (2013), which became determinant for RRI. Their approach distinguishes four dimensions:

– *Anticipation*: Anticipation means thinking ahead of possible event sequences (what if?) and looking at strategic action.
– *Reflexivity*: Reflexivity means a rethinking of one's own (moral) position and refers to individuals as well as institutions;
– *Inclusion*: Inclusion refers to open, pro-active cooperation with other social actors;
– *Responsiveness*: Responsiveness refers to the ability to remain open to change, especially in order to be able to correct possible wrong decisions.

COST: A Case of RRI?

The RRI approach is located within the context of a general debate about opening science to social issues and valuing non-formal, societal knowledge, which started at the latest with the discussion about environmental protection in the 1980s. A coproduction of knowledge was demanded (cf. Callon, 1999; Mieg & Evetts, 2018), similarly understood as "mode-2" science (Gibbons et al., 1994) or transdisciplinarity (Scholz, 2013). Since the inception of mode-2, which has mainly focused on the relationship between academia and society, and the evolving role of universities as a mediator of that relationship, various additions and updates have been made to the idea. Mode-2 similarly acknowledges the importance of knowledge produced in society or the "agora" (Nowotny et al., 2001). These approaches include engaged scholarship (Van de Ven, 2007; Van de Ven & Johnson, 2006), and collaborative research initiatives between academics and practitioners (Ren & Bartunek, 2020; Sharma & Bansal 2020; Rynes et al., 2001).

In this light, we examine an established EU research initiative (COST), and a current project focusing citizens' knowledge. The aim is to assess the contribution made by such a project in terms of the responsibility of science for society. COST (European Cooperation in Science and Technology) is a funding organization founded in 1971. Through projects called COST Actions, it offers scientists in Europe the opportunity to hold joint conferences, network, and publish together. COST further elaborates the network idea as follows:

> COST is bottom up, this means that researchers can create a network—based on their own research interests and ideas—by submitting a proposal to the COST Open Call. The proposal can be in any science field. COST Actions are highly interdisciplinary and open. It is possible to join ongoing Actions, which therefore keep expanding over the funding period of four years. They are multi-stakeholder, often involving the private sector, policymakers as well as civil society. (COST, 2021: https://www.cost.eu/about/about-cost/)

The main success of COST projects lies—according to their own statements—not only in scientific networking but also in building bridges to society and linking up with societal concerns. Thus, COST's impacts can be defined as (COST, 2019):

- *Scientific impact*: Interdisciplinary collaborations leading to breakthrough science;
- *Societal impact*: Bridging the innovation divide and participation gaps, and enabling skilled labor mobility and networking throughout the European Research Area.

Inclusion and Coproduction of Knowledge in Placemaking in Europe

In the following we present the COST project "Dynamics of placemaking and digitization in Europe's cities" (CA18204). This project includes participants from 34 countries with diverse disciplinary backgrounds (see Box 12.1). The project demonstrates how inclusion and coproduction of knowledge can redefine the role of research. Here we present the project based on four characteristics: (1) placemaking, (2) heritage, (3) civic engagement, and (4) digital culture.

Box 12.1: Dynamics of Placemaking and Digitization in Europe's Cities (COST Action CA18204, 2019–2023)

This Action will investigate how placemaking activities, like public art, civil urban design, and local knowledge production reshape and reinvent public space, and improve citizens' involvement in urban planning and urban design. Placemaking implies the multiplication and fragmentation of agents shaping the public realm. The Action aims to empower citizens to contribute by means of citizen knowledge, digitization, and placemaking to diverse ways of interpreting local identities in European cities. The added value of digitization—understood here basically as the ongoing process of converting any kind of data from an analog into a digital format—will be analyzed through the ways in which it impacts urban placemaking processes of local communities (the project explicitly addresses digitization beyond smart city concepts). Studying urban placemaking and digital practices of various local communities throughout Europe's cities, this Action will understand and analyze:

- The impact of digitization on the common placemaking practices of urban local communities;
- The changing processes of citizens' local knowledge production of placemaking;
- The influence of digitization on the governmentality of local neighborhoods and co-creation of public space by various societal actors.

Drawing on recent theoretical insights that point to the importance of placemaking, widening citizen knowledge, and broader application of digitization and digital communication, the Action seeks to develop new methods for studying and comparing the effects of disseminating local urban knowledge beyond cultural and societal borders. By doing so, it develops European urban research both theoretically and methodologically, finding ways of channeling the results into the wider urban planning and governance processes.

The introduced cases are:

- Housing estates in Accra and Douala
- Living Memorial in Budapest
- Memorial Mapping of Violence Against Women in Europe
- Haydarpaşa train station in Istanbul
- Lost heritage in Tirana
- Port cities in Europe
- Russian Tsar neighborhood in Ljubljana

The project duration is four years (2019–2023).

Placemaking

This specific COST Action focusses on placemaking. Placemaking has become one of the central concepts of designing urban environments and a powerful people-based instrument for intervening in urban development processes at the local scale (Fürst et al., 2004). Three major aspects may be highlighted. First, placemaking is a **locally** determined response by planners, politicians, and people to the phenomena and processes of globalization, in which local and intercultural knowledge and local particularities are expressed as a counterpart to global interdependence. Second, placemaking as a **participatory** strength in urban development aims at preserving cultural identities and promotes bottom-up processes of urban renewal and urban design. Third, placemaking as an agent for inclusion is primarily in conflict with economic processes of displacement, gentrification, and differentiation, and is thereby situated in the context of a **socially sustainable** city (Dupre, 2019).

Placemaking, in its critical normative dimension, claims that local practices and local knowledge are the most sustainable approaches for development processes at the local scale. Placemaking may offer a critique of current urban development practices, consequently facilitating a change in the governance of urban development in general, namely towards bottom-up, participatory, community-led processes (Drilling & Schnur, 2019).

Heritage

Heritage and heritage sites are often highlighted within placemaking processes, and thus all of the introduced case studies of the COST Action on dynamics of place-making and citizens' knowledge deal with heritage and, more specifically, with heritage in the context of urban planning. The disciplines of heritage and urban planning emerged in response to the massive and rapid uncontrolled urbanization associated with industrialization of Europe's cities in the nineteenth century. Since then, heritage and urban planning share some approaches when considering single monuments, but also contrasts through their positions on careful enhancement of urban structures in heritage conservation versus tabula rasa approaches seen in planning. At the beginning of the twentieth century both were scientific, rationalized fields of expert knowledge that became part of local politics through formalized administrative procedures, but the situation changed slightly in the second half of the twentieth century. Starting with the **rediscovery of the historic city** as places to live and work in the 1970s, communities requested greater involvement in decision- and place- making (cf. Choay, 1992; Hosagrahar, 2017). There is an ongoing **critical debate** in heritage studies that challenges the established knowledge on monuments and heritage conservation. The debate even challenges ideas of what constitutes heritage, who defines it, and how and by whom historic site, memory, and memorization are constituted, performed, and (re-)framed (cf. Smith, 2006; Waterton & Watson, 2015). Agents and **access** have become predominant issues in this discussion (Oevermann & Gantner, 2019), and have broadened the scope of research into how citizens' knowledge contributes to science and practices.

Civic Engagement

In terms of heritage concerns within urban planning, the role of science is blurred between theory and practice, and generally includes diverse constellations of actors and varying perspectives. The contribution of civic engagement is obvious and increasingly acknowledged as expertise in this transdisciplinary field. Institutional and social innovation are ongoing (although not in all places or countries), and is necessary to conceptualize and implement locally appropriate forms of sustainable urban development (cf. Mieg & Töpfer, 2013). The focus of RRI and this COST Action is less the market-driven innovation problematized by von Schomberg (2013) but rather the societal impact of knowledge as highlighted by the Lund Declaration (2009, 2015).

A core issue of today's placemaking is the application of two sides of knowledge: **research 'from above,'** namely looking into planning regulations and interviewing city planners, politicians, and other authorities and specialists, and **research 'from below,'** such as phenomenological studies of places at eye-level, interviews with local citizens, and more. In this sense, a mixed methods approach (cf. Bryman, 2008; Hesse-Biber, 2010) is often beneficial, combining knowledge from both quantitative sources (e.g., statistics) and qualitative sources (e.g., interviews, questionnaires, fields studies, observations), since:

> There is no single best method—questionnaire, interview, simulation, or experiment—for studying people's adaptions to their environments. One chooses methods to suit the problem and the people and not vice versa. These methods are generally complementary rather than mutually exclusive (Sommer, 2007, p. 221).

Thus, the production of knowledge in research is broad; as such, many perspectives (including contrasting perspectives) of a case allow a better and deeper understanding of its complexities. As a result, methodological approaches are manifold (see Box 12.2).

Box 12.2: Dialogical Research

Various approaches born out of social interactionism, dialectical theory, and dialogism, which are grounded in such fields of organization studies, including knowledge management, have offered various views of dialogical production (Tsoukas 2009a, b, 2019). Further methodological approaches, and specifically ethnography, emphasize breaking down the distinction between the observer and the observed, and aim for dialogical co-production of knowledge in order to deal with the complexity of urban environments in research (cf. Stoller, 1989, 2009; Shah, 2017).

Digital Culture

The emergence of digital culture and the ubiquity of digital technologies have added an additional layer to placemaking, especially in the neighborhood context. With neighborhood being a place-based concept, the digital and the analogue space are always interrelated to each other, leading to the emergence of hybrid spaces and **hybrid networks** (Jonuschat, 2012). Neighborhood platforms, apps, social media, and community websites are increasingly used by residents to address local concerns, to create and extend dense networks of weak ties that facilitate action, and to gain and provide access to resources and information for action or **mobilization** at the local scale. Networks and resources can be mobilized sporadically and at short notice for placemaking activities, such as cleaning public spaces together, mobilizing for a demonstration against top-down-initiated activities in the neighborhood, but also for long-term projects such as urban gardening and collective actions to reactivate public space (Johnson & Halegoua, 2014).

The use of digital tools may enhance the visibility and publicness of residents, their concerns, and placemaking activities, which, in turn, can increase their impact on politicians or create an effective counter-public and, thereby, create opportunities to co-create urban space. The use of digital tools in the neighborhood context can also foster **a sense of belonging** and strengthen place identity. It can help to create powerful collective narratives, imageries, and representations of neighborhoods and, thereby, produce a sense of community (Menezes, 2019). Nevertheless, digitalization also raises questions of in- and exclusion. In a society of access, where being connected and having access is crucial, pre-existing inequalities and segregation can be perpetuated or even exacerbated (Rifkin, 2014).

Focus of Research

To sum up, we can identify four main domains of science for heritage case studies:

1. The initial contacts of planners, politics, and investors with citizens and citizens' organizations;
2. Creating and/or supporting processes and dynamics of engagement, ideas, and visions;
3. Collecting, analyzing, systematizing, and evaluation of data, including an open debate on the production and results of citizens' knowledge,
4. As well as the transmission of outcomes to all involved.

The following two subchapters introduce the heritage case studies, which either contrast, differentiate and add to existing knowledge and practices in placemaking through specific initiatives, or else facilitate establishing common ground within a wider constellation of societal actors and both, as we see, contribute in different ways to responsible research.

Citizens' Knowledge: Contrasting, Differentiating, and Adding Perspectives

This section introduces and reflects and cases that contrast, differentiate, or add knowledge to already established definitions and uses of heritage. Section 12.3.1 explains that heritage is not only about present concepts of definition and use, but is also constituted by its historic dimension. Section 12.3.2 discusses three cases that contrast top-down decisions of heritage and placemaking. Section 12.3.3 introduces the ways in which lost built heritage can be substituted through citizens' narrations and digital tools. In Sect. 12.3.4 these cases are reflected along the four criteria of RRI defined by Stilgoe et al. (2013), and an overview is provided on modes of implementation.

The Historic Dimension of Heritage: The Case of Housing Estates in Sub-Saharan African Cities

Citizens' knowledge and the historic dimension of heritage are discussed in the case of housing estates in Nairobi, Accra, and Douala, among others. The case shows how citizens' knowledge improves the understanding of this historic dimension regarding built forms and typologies, use, informal adaptations over time, and the current state of the art.

Native Housing Estates in Sub-Saharan African Cities

This case study critically compares a series of little-researched, twentieth century publicly commissioned 'native' housing estates in Nairobi (Kenya), Accra (Ghana), and Douala (Cameroon) as of the 1920s. Local and/or citizens knowledge was and still is produced in various ways and is crucial for insights into the estates' forms, typologies, and adaptations over time. The ANT-related method (actor-network-theory-related method), as applied in this research, intends to identify the human and non-human actors involved in the planning and design of the estates, as well as in their subsequent (often informal) adaptations (Martin & Bezemer, 2020). Their complex and ongoing interaction has resulted in hybrid estates wherein global (here meaning Western), colonial models merged with/have been adapted to local native housing typologies, dwelling rituals, land ownership rules, and to societal and economic changes over time. Most estates are now at risk of replacement and/or gentrification, partly due to lack of local and historical awareness. Due to a lack of formal sources regarding planning documents, residents are involved in in-situ fieldwork, including mapping practices and interviews, initiated by a young researcher as part of her PhD research. As such, local citizens are invited to contribute their knowledge on issues related to historical awareness, placemaking, identification, and place-boundness. A crucial side-effect is that citizens' knowledge allows for under-standings of the estates' value—such as in ethnic, emotional, historical, and material terms—which, in turn, might influence heritage and conservation matters.

In contrast, local knowledge was only rarely sought as a source for the original design and production of public 'native' housing estates in Sub-Saharan Africa; especially under colonial regimes, its role and transfer were unforeseen side effects of the complex interplay of actors involved in shaping the housing estates (Ese, 2014; Ese & Ese, 2020).

Three findings can be highlighted: first, citizens' knowledge is traceable in the way that 'native' dwelling habits and building practices prompted the transforma-tion and mutation of global—including colonial—urban models and dwelling typologies used in the design and planning of these housing estates. Second, citi-zens' involvement incited the estates' subsequent non-legal adaptations and transfor-mations. Third, the estates' residents still participate in local knowledge production via their active role in the digital mapping of their estates' current situation and conditions; in doing so they draw on their own experiences, symbolic-physical land-marks, surviving stories, and so on. Potentially, conservation and development paths alternative to gentrification can be established that build on this citizen knowledge.

Contesting and Enriching Perspective: The Cases of the Living Memorial, the Memorial Mapping of Violence Against Women, and Haydarpaşa Train Station

Citizen knowledge and contesting top-down memorization is the focus of the Living Memorial and the Memorial Mapping of Violence Against Women (both Budapest); the enriching potential of citizens' knowledge is illustrated through the case of Haydarpaşa Train Station (Istanbul). The three cases show how people contest top-down, established imagination, narratives, and memorization, and thereby broaden the cultural heritage landscape in Europe though citizen knowledge.

Living Memorial, Budapest

The Memorial to the Victims of the German Occupation became the object of heated discussions for various reasons. On the one hand, the memorial got inaugurated in a hurry without any kind of public dialogues: from the date of announcing the govern-ment's plan of erecting a memorial to its actual realization only six months have passed. On the other hand, its "message" articulating that Hungary was the victim of Germany—thus also shifting the responsibility over the Holocaust to the German authorities—was also found unacceptable by many (Kunt et al., 2017).

The appearance of the living memorial was closely connected to eight actions and communities created on Facebook, most importantly *"Holocaust and my Family"* and *"Living Memorial."* In both groups, knowledge production activities were crucial: personal memories of the Holocaust—many of which had never previously been narrated—started to flood the digital space, and were later read aloud in theatres and then also published in a book, thus further strengthening the aspect of knowledge sharing. At the same time, these happenings also reinforced the physical appearance of a number of practices at the very site of the Memorial to the Victims of German Occupation in Budapest. As a result of these guerrilla actions, Living Memorial was spontaneously created from the personal objects and messages of people protesting, which currently serves as a counter-monument to the Memorial to the Victims of German Occupation. This not only symbolizes a bottom up approach to placemaking activities vis-à-vis the top-down process of erecting the official memorial, but also represents an approach that argues for the importance of citizens' involvement in commemorative projects.

From a critical perspective, an action might struggle with how to exert influence outside the frames of particular projects, and sometimes also how to maintain a practice at the site of a project in the long-term, but is responsible in the sense that citizens articulate that living with the past has serious social and political implications. Here, digital space clearly served as a primary field of protest, but what is interesting is how digital practices became connected to actual placemaking activities in the urban space.

Memorial Mapping of Violence Against Women

A similar contesting placemaking is the ongoing project of the Memorial Mapping of Violence Against Women (based on the planned erection of a memorial in Budapest dedicated to the "Memory of Rape in Wartimes: Women as Victims of Sexual Violence": Elhallgatva, n.d.). These 'performative monuments' memorialize violence against women and gender-related traumata in public space in Europe. In the context of the Budapest case study, the Memory of Rape in Wartimes promise to be an exercise in the democratization of placemaking: the democratic awareness of how public memorials should be conceived and realized in the framework of a wide public discourse and acceptance. In contrast to the current Hungarian government's cynical and neglectful practice, which is imbued with revisionist and nationalistic narratives of some imagined 'heroes' of Hungarian history carved in nineteenth-century-style sculptures and erected almost 'overnight' without any public, scientific, or expert consultation and debate. In contrast, the Memorial Project includes citizens' knowledge through oral historical interviews and documents from private archives, and counteracts dominant xenophobic, gender-hostile, nationalistic narratives.

Citizens' knowledge on this special issue of rape in wartime is especially relevant, since it is rarely documented in historical archives. Within the framework of the COST Action the interest in this particular case is (at least) twofold: not only is the topic newly raised in the public discourse, but it also empowers citizens to contribute with their own history and memory to a process, when the realization of a memorial becomes traceable in public debates, in public lectures, in a transparent procedure of the applications and selection process of artists.

The Memorial Mapping project will create a digital map that will not only show these memorials throughout Europe, but also indicate the process through which they have been realized. The research aim is to digitally visualize, detect, analyze, and systematically map these forms of placemaking as 'performative monuments' in Europe. Furthermore, it highlights the omission of this issue from urban space— whether in the sense of failed attempts or to-be-realized projects. These mappings will also include 'collateral' events, such as performances, actions, demonstrations, etc. in and around the memorials, and focus on the interactivity between the monument and the spectator.

Haydarpaşa Train Station

Haydarpaşa Train Station in Istanbul is a site demonstrating the contestations in recent urban transformation and citizens' involvement to become core actors in the process. The station, constructed in 1908, is an iconic building of the early twentieth century and served as one of Istanbul's two main centers of water and transportation infrastructure. The station was closed during the last decade, amid speculation that it would be turned into a hotel and shopping mall. It was also defined as cultural heritage by local stakeholders and the community, and inspired ongoing public protests at the site. The multiple ways to narrate and reimagine heritage, in addition to the dialogical intersection of the people, the site, and the practices, reveal a shifting phenomenon of place, site, and memory. Located each time differently in spatial and temporal contexts and cultural values, Haydarpaşa Train Station showcases the citizens' knowledge in reimagining and redefining cultural heritage, negotiating formal and informal ways of making heritage and place.

In this sense, many events organized by the community introduced dynamic understandings of space: a book festival, advertised with narrations of the sea and the animals that made up this territory; a social media channel for sharing and narrating the past and the present of the site, referencing it as a site of memory across generations; marches, picnic, and dance events at weekends; identifying the site as a graveyard if not used for its original function; and opposition to the musealization of the space, imagining this cultural heritage site as lived space. All referred to the urban, social, and cultural importance that citizens attach to the site. The citizens' involvement produced a lively period for this heritage site after 2012, and introduced "a dynamic understanding of space that no longer appears as a fixed entity but depends on how it is visited, used and dealt with" (Haldrup & Bærenholdt, 2015, pp. 54–55). The Haydarpaşa case demonstrates "heritage as practices and performed, subjective and situational and emergent in particular settings" (Haldrup & Bærenholdt, 2015, p. 52), hence open-ended and site-specific, therefore with potential for criticism.

COST Action provides the possibility of creating environments of participatory dialogues among multiple actors—community, governments, NGOs, public and private sectors, academics, creatives, and more—to better understand each other, to understand the perceived contradictions and controversies, and the possibility of empowering citizens' knowledge in contributing to the process of 'making' the future.

Citizens' Knowledge as Substitution for Lost Heritage: The Case of Tirana

The Tirana case demonstrates how citizen knowledge can help to substitute for lost heritage, with digital redefinition of the city's communist-era built heritage (see Table 12.1).

Table 12.1 Citizens' knowledge: contrasting, differentiating, and adding perspectives and the four dimensions of responsible research

Dimensions	Cases				
	Housing estates in Accra and Douala	Living memorial, Budapest	Memorial mapping of violence, Europe	Haydarpaşa train station, Istanbul	Tirana's lost heritage
Anticipation	Mapping of alternative heritage values as basis for community-led conservation and development	Chronicling the lives of civilian victims of war and occupation	Showing a barely visible dimension of life in wartime	Reimagining and redefining cultural heritage	Substitution of lost built heritage through memories, experiences, and stories
Reflexivity	–	–	–	–	–
Inclusion	Residents as experts of the heritage (e.g., oral history, mapping the transitional, informal)	Citizens' private archives and knowledge as part of the history (mapping) & collateral events, blending analogue and digital spheres	Citizens' private archives and knowledge as part of the history (mapping) & collateral events, blending analogue and digital spheres	Reuse of cultural heritage by citizens	Mapping, games, application for accessing, interacting, and contributing
Responsiveness	Citizens' knowledge as part of heritage conservation; countering potential loss through private development	–	–	–	–

Lost heritage in Tirana

After the fall of communism, Tirana underwent very dramatic dynamic urban mor-
phological and aesthetical change. In the new capitalist economic system, the urban
texture of the city is densified at the expenses of public and green spaces.
Furthermore, the capitalist system has notably altered the sense of place, from a city
characterized by mid-rise buildings to one that has rapidly added high-rise
developments.

Such turbulent urban development is leading to the demolition of cities' older
and historical physical pattern (Manahasa & Manahasa, 2020). In many cases, pressure from developers shows no appreciation even for those buildings of great importance to a city's collective memory or those of outstanding architectural or historical
value. To date, urban villas from the pre-socialist period (pre-1944) are still being
lost daily, replaced by high-rise apartment complexes.

In this context, the digitization of knowledge, especially that of citizens who
experienced the socialist period, is used as an important source—firstly to register
memories and stories about the lost buildings and landmarks or atmospheres of the
demolished and transformed streets and neighborhoods. Beyond old images, the
digitization of personal experiences regarding the lost city's physical component is
also very important. Due to the very specific context of Tirana, whose post-socialist
physical urban texture is very different from the socialist period, digital mapping is
used to re-locate lost landmarks, gathering neighborhood locations and measuring
nostalgia for cardinal urban spaces such as the city's main boulevard.

The COST Action uses several tools to encourage citizens to contribute with
their knowledge, using first those methods that aim to stimulate the citizen involvement and participation with a series of workshops, community meetings, or open
lectures regarding specific themes targeting a particular audience. Secondly, beyond
digitizing and recording the collected data, methods such as mapping, games, applications, and programs are combined to reassemble memory, reflection, will, desire,
or other contributions to the placemaking process. At the same time, various technological means facilitate access, enable joyful interaction, and add diverse con-
tributory tools to the placemaking process.

Reflecting the Four Dimensions of Responsible Research

Placemaking and the introduced heritage cases highlight some aspects within the
four dimensions presented by Stilgoe et al. (2013) that seem most relevant to understanding responsible research in this field: *Anticipation* includes consideration of
alternatives and strategic actions that allow both the integration of complexity and
different perspectives. *Reflexivity*, namely the rethinking of one's own (moral) position, is necessary to take citizens' engagement seriously and to establish a dialogue
as equals. *Inclusion* refers to open, pro-active cooperation with other social actors
and is a prerequisite for access to heritage and heritage sites. *Responsiveness* refers

to the ability to remain open to change but also include consideration of long-term resources, thus the initiative and running of projects can become facilitated. Mutual learning is part of this responsiveness.

As Table 12.1 shows, inclusion is realized in all of the projects presented. What is surprising is that we see anticipation implemented in all cases, and their significant contributions to making visible and accessible further, alternative dimensions of heritage. This anticipation thus allows for access, engagement, and inclusion, with the process of mapping being especially supportive. Responsiveness seem to be implemented only in one case, and functions as part of a risk management strategy against the loss of heritage through unconstrained redevelopment. Reflexivity is less considered or contingent as a component of RRI.

Citizens' Knowledge: Developing Common Perspectives

For placemaking projects the involvement of citizens is key (cf. Ellery & Ellery, 2019; Urbact, 2019). In this respect, it is relevant to think about who has been involved (using criteria such as gender, age, occupation, and others); how many people have been involved; whether it is possible to regard them as being representative of the wider community, or whether their knowledge should be interpreted and included in the project as mere insights; and how they have been involved (Kvale, 2008). This is possible to analyze with several follow-up interviews, or through questionnaires, workshops or observations or other appropriate research methods. Placemaking planning-processes require common ground among a broad constellation of actors in order to achieve societal acceptance. Section 12.4.1 introduces two cases, namely port cities and the Russian Tsar neighborhood of Ljubljana, Slovenia. The cases show how different forms of knowledge, including citizen knowledge, are integrated and—despite the challenges of such work—achieve consensus. Section 12.4.2 reflects on the two cases in accordance with the four criteria of RRI defined by Stilgoe (2013), and provides an overview of modes of implementations with respect to Sect. 12.3.4.

Common Ground: Historic Port Areas of Europe and the Russian Tsar Neighborhood of Ljubljana

The degree to which placemaking activities around heritage take place in European port areas depends largely on the long-standing relationship between local port and city actors. Common ground is needed for further conservation and development. Also, in the case of Ljubljana, consensus was reached after establishing dialogues that give respect to citizens' perspectives.

Historic Port Areas of Europe

The importance of ports and the existence of port authorities as economically important and powerful actors makes placemaking around and in ports a particularly challenging topic. In this respect, many different data have to be managed; the Waterwheel methodology is a suggestion for systematization of such tasks.

Citizen involvement in the planning of (former) port areas varies extensively from between cities and countries; nevertheless, ports are often large industrial areas under the control of largely independent port authorities with their own planning rights. Their primary goal is to facilitate the transport and transportation of goods and people, and the needs of the locality are often secondary. Placemaking activities often only occur in historic port areas around selected heritage sites that are no longer used for port activities. Placemaking of port heritage sites can contribute to the creation of maritime mindsets, i.e., an awareness of the particularities of port cities, and therewith also contribute to future placemaking activities in active ports.

RRI can help evaluate programs that assess the environmental sustainability, and societal desirability of an innovation by considering input from multiple stakeholders, including in historic and contemporary port areas (von Schomberg, 2011; see Table 12.2). Transparent communication within these evaluations helps the various innovators in the built environment—governments, urban planners, and industry representatives, and the general public—to optimize current planning, making it more reliable and acceptable to society and reducing its risks. Understanding the inherent qualities and path dependencies in port/city relationships through deep-mapping can facilitate future planning. Through an agile methodology of exploring and analyzing historical sources, this approach seeks to show how different spatial and institutional frameworks facilitate or hinder collaboration collaboration among local actors and the efficacy of placemaking (Hein & van Mil, 2020). The outcome is organized as a datawheel that brings together and makes accessible the knowledge produced on multiple levels, such as policy making, branding, funding, media, etc., and consisting of varying perspectives.

Table 12.2 Citizens' knowledge: Developing common perspectives, and the four dimensions of responsible research

	Cases	
Dimensions	Historic Port Areas	Russian Tsar neighborhood
Anticipation	Understanding path dependencies	Shared vision of heritage-based urban regeneration vs. top-down planning
Reflexivity	Deep-mapping as a tool to reflect and better understand collaboration	Rethinking the participatory process (new tools, both analogue and digital)
Inclusion	Building a broad consensus	Coalition of self-organized, citizens' placemaking projects
Responsiveness	Waterwheel as data management tool	Established, heterogeneous, local citizens' network in cooperation with institutions

Various projects—such as the Elbphilharmonie in Hamburg and the M4H in Rotterdam (Rotterdam Makers District)—analyze and continue historical place-making practices which serve as an orientation for the ongoing structuring of (historical) information and the analysis of established relationships among different actors as a foundation for design. Such understanding can help facilitate citizen involvement, knowledge development and engagement, and also present an opportunity to emphasize the role of imaginaries and narratives for the development of shared values. This approach can also facilitate the development of other carefully promoted placemaking projects at the intersection of land and water. At a time of climate-induce sea-level rise, it is important to acknowledge and collectively plan for water as a continuous element. The proposal of a datawheel has therefore been explored as a Waterwheel by a group of researchers at TU Delft working on digital humanities (Hein et al., 2020).

The Russian Tsar Neighborhood, Ljubljana

The Russian Tsar neighborhood of Ljubljana, Slovenia, is mainly influenced by a major socio-economic change from the former socialist society with a planned economy, to a capitalist society with an open market economy, including rapid motorization, the dream of a suburban detached house with garden, shopping malls, and neglected, diminished, or privatized public areas (Nikšič et al., 2018). The current rigid, top-down-oriented urban planning system in Slovenia still develops urban regeneration strategies within the rather closed professional circles as part of the briefing of strategic development goals. At the same time civil society is activating itself to address the most pressing problems within the neighborhoods. Citizens self-organize to improve their living environments through volunteered activities, very often in the form of community-led placemaking and tactical urbanism. While these activities have immediate and clearly positive short-term effects resulting in improved living conditions in some particular locations, no major long-term or systemic influence is achieved for better living in the aged housing estates, also because civil initiatives lack any systematic support and thus eventually lose momentum (Nikšič, 2018).

In such conditions it is essential that actors start to cooperate: The local knowledge that exists and is performed through bottom-up activities must become recognized by official urban planning actors; at the same time, the readiness of the civil initiatives to cooperate with the planning system must also increase in order to achieve any breakthrough and enable the two sides to benefit from their joint action and cooperation. This approach is far from easy in practice, which was well reflected in the case study of urban regeneration endeavors in the Russian Tsar neighborhood of Ljubljana during the Human Cities project that ran from 2014 to 2018 (Cité Du Design & Clear Village, 2018). The residents were not only seen as partners in the process from the very start but were also given the role of the best local experts with the most precise insights into the state of the neighborhood, its potentials, as well as obstacles to qualitative improvements. Citizens pointed out both assets as well as problems of the neighborhood that were not mapped by the business-as-usual-analyses conducted by the urban planning office. This was only possible through the development of a series of get-together activities in which citizens and professionals spent time together within the neighborhood and talked and listened to each other via various communication tools, such as neighbors' walks, local roundtables, public picnic, hands-on workshops, and exhibitions. Only some of the classical participatory tools were used (on site, face to face interactions) and these were almost always attended by the same groups of residents. Therefore, secondly, the digital tool *Photostory* was tailored to allow residents to pass on their knowledge through photography and their captions within various categories, such as 'My neighbor' to seek local insights into who constitutes the local community in the eyes of the community itself; 'Professions in my neighborhood' to reveal the problem of the existing and absent services in the neighborhood; and 'Shared values of my neighborhood' to reveal the intangible cohesive elements that contribute to the notion of the community. This digital tool managed to attract new groups of residents to contribute their knowledge, ideas, and insights to the common debate, and new understanding of the place was also gained by the official urban planning actors.

Reflecting the Four Dimensions of Responsible Research

Both case studies illustrate the importance of matching citizens, politics, and administration, and of defining a common perspective, in order to contribute to all four dimensions of responsible research. Unlike the cases presented in Tables 12.1 and 12.2 shows how reflexivity and responsiveness also contribute in these two common-ground cases. Ethical engagement with citizens and communities seems to be paramount in that such groups are involved in research projects and initiatives as partners from the start. Engaged scholarship advocates for a strategy of arbitrage: rather than urban design being presented to communities, communities are instead worked with from the start to improve the final outcome. In both cases, the projects are not 'done to' communities but instead are 'partnered with' communities for better outcomes. Science and politics both have a role in bringing into play the citizens' knowledge over the long term. The presented case studies exemplify the objectives of the Science with and for Society (SwafS) program.

Speaking normatively, the safeguarding of responsibility, on the one hand, depends on guaranteeing certain performance criteria and procedures, in particular the transparency with which the entire placemaking process is conducted. Ranging from the transmission of the objectives of the action, to communication and information, and the use that is made up of the knowledge produced, that is, it always has a political component. On the other hand, during the process and respective completion of each of its phases and a given objective or product, it is important to find ways to guarantee citizens some type of social control over the ways in which knowledge produced is used. In this sense, researchers have a responsibility to safeguard the appropriate status of citizens' knowledge and to address issues of communication and transparency.

This may demonstrate the need to safeguard an arena of debate and decision that, from a democratic point of view, safeguards interests, impacts, and risks in a considered way.

Reflexivity is fundamental, namely to evaluate procedures, achievements, and challenges, allowing fine-tuning of subsequent phases, actions, or procedures, as well as approaches to producing knowledge (theoretical and more scientific). That means reflectivity can be related the notions of shared authorship or shared knowledge. This principle depends on the context of the placemaking and the underlying actions, the particular stage of the process, the way the process is outlined and streamlined, and on the specificities and characteristics of social actors involved in the process.

The Role of Science: From Coproduction of Knowledge to Responsible Science

Our reflection on placemaking projects in light of RRI (Stilgoe et al., 2013) yielded two unexpected findings. Firstly, that inclusion and anticipation occur together. Anticipation appeared here in the form of alternative perspectives, and seems relevant—although the projects were not specifically about innovation (as presupposed by RRI). Inclusion, in turn, was often implemented through citizen science: citizens doing research. Thus, citizen science as a success story of SwafS (European Commission, 2020) is also visible in placemaking. This becomes obvious in the case studies where citizens' knowledge contrasts, differentiates, and adds perspectives on heritage issues. The case of the Memorial Mapping of Violence Against Women and its approach to using oral history and open private archives provides a good illustration of the importance of these specific research tasks. We can argue that, in our context of planning and heritage, *anticipation of alternative perspectives is conducive to inclusion*. In our cases, the coproduction of knowledge (as required by SwafS) seems to work only with anticipation. This is evidenced by our first five placemaking cases.

The second finding concerned *reflexivity*. Reflexivity occurred when placemaking projects aimed to develop common ground in planning practices. Even though all the COST Action projects demonstrated that—and showed how—citizens' knowledge contributes to science and allow for inclusion, further dimensions of RRI as defined by Stilgoe et al. (2013), such as reflexivity or responsiveness, were not addressed. However, cases that allow for developing common ground in planning practices might not only involve reflexivity but also better contribute to *responsiveness* in gaining institutional stability, be it through the acknowledgment of citizens as partners or specific analytical and data management tools as the Waterwheel. We can even assume that the reflection of different perspectives develops an ethical quality, and from there promotes responsible action (cf. Mieg, 2015).

In addition, *digital* tools help to bridge participation gaps, encourage brain circulation within the ERA, and thus have a societal impact (even beyond smart city concepts). The cases show that digital tools broaden access to and engagement with processes of defining, using, and conserving heritage through collecting, systematizing, and presenting findings, and in this way stimulate anticipation and support inclusion. The use of digital interfaces, digital tools, online platforms, or social media channels may create different engagement channels that enhance processes of local networking, exchange, discussion, community learning, and action, and thereby allow for a citizen-centric approach. However, as already mentioned, those groups who lack access to digital technology or knowledge about how to use digital tools are at risk of exclusion from such processes and opportunities.

What do we learn from the COST project, from the perspective of the responsibility of science? The cooperation between research with society, as it has been presented as mode-2 research (Gibbons et al., 1994), has evolved. It is not simply about exchange and collaboration, but more about *"mutual learning,"* on the part of citizens as well as science, in the sense of the transdisciplinary approach (cf. Scholz, 2013). Science—and scientists—can learn to think along other dimensions in planning and dealing with heritage. Here, *imaginaries* (cf. Strauss, 2006) play a new role, not as cultural beliefs to be studied but as productive research and design tools in placemaking, because it is also about the power to define local development (cf. Jasanoff & Kim, 2015), see for example the case of the lost heritage of Tirana.

Responsible science seems particularly relevant for dealing with *heritage*. We see that citizens' knowledge contributes in different ways to heritage studies and heritage practices. Citizens' knowledge improves the understanding of the historic dimension of the heritage and the sites, e.g., the history of uses by people; this knowledge contests and contrasts governmental memorization, and may compensate—to some extent—for lost heritage through narration and digital tools. These contributions of citizens' knowledge are needed to understand the complexity of heritage, sites, and placemaking. The high value of these added perspectives is the real benefit of making heritage dimensions visible and accessible. This mode of anticipation, as the cases in Sect. 12.3 show, allows the mobilization of people, to engage and to become included. In the field of heritage and placemaking citizens' knowledge is of the greatest value, because they know what outsiders, professionals, or politicians do not know—or are unwilling to recognize.

Furthermore, citizens' knowledge adds to formal planning processes and enables heritage identification and acceptance of planning procedures. The cases in Sect. 12.4 show that the inclusion of citizens' knowledge is challenging and requires the establishment of a dialogue that takes citizens seriously—only then may acceptance and consensus be possible. Additionally, these cases implement modes of reflexivity and responsiveness, through which the uses of heritage can be balanced over the long term, within a changing society and uncertain future.

Acknowledgement This publication is based upon work from COST Action CA18204, supported by COST (European Cooperation in Science and Technology).

References

Bryman, A. (2008). *Social research methods* (3rd ed.). Oxford University Press.
Callon, M. (1999). The role of lay people in the production and dissemination of scientific knowledge. *Science, Technology and Society, 4*(1), 81–94.
Cité Du Design, & Clean Village (Eds.) (2018). *Challenging the city scale: Journeys in people-centred design*. Birkhauser.
Choay, F. (1992). *L'allegorie du patrimoine*. Editions du Seuil.
COST. (2019). *Impact of networking*. COST Association. https://www.cost.eu/wp-content/uploads/2020/02/COST_ImpactBrochure_7_WEB_1P.pdf
COST. (2021). *About COST*. https://www.cost.eu/who-we-are/about-cost/
Drilling, M., & Schnur, O. (2019). Neighborhood development. In A. Orum (Ed.), *The Wiley-Blackwell encyclopedia of urban and regional studies*. Wiley-Blackwell. https://onlinelibrary.wiley.com/doi/full/10.1002/9781118568446.eurs0215
Dupre, K. (2019). Trends and gaps in place-making in the context of urban development and tourism: 25 years of literature review. *Journal of Place Management and Development, 12*(1), 102–120.
Elhallgatva (n.d.). *Elhallgatva [Silenced]: Memory of rape in wartimes: Women as victims of sexual violence*. Retrieved February 24, 2021, from, https://www.elhallgatva.hu/?lang=en.
Ellery, P., & Ellery, J. (2019). Strengthening community sense of place through placemaking. *Urban Planning, 4*(2), 237–248. https://doi.org/10.17645/up.v4i2.2004
Ese, A. (2014). *Uncovering the urban unknown: Mapping methods in popular settlements in Nairobi*. PhD thesis. The Oslo Scholl of Architecture and Design.
Ese, A., & Ese, K. (2020). *The city makers of Nairobi*. Taylor & Francis.
European Commission. (2012). *Responsible research and innovation: Europe's ability to respond to societal challenges*. https://ec.europa.eu/research/swafs/pdf/pub_rri/KI0214595ENC.pdf
European Commission. (2020). *Science with and for society in horizon 2020: Achievements and recommendations for Horizon Europe*. https://op.europa.eu/en/publication-detail/-/publication/770d9270-cbc7-11ea-adf7-01aa75ed7
European Commission. (2021). *Science with and for society*. https://ec.europa.eu/programmes/horizon2020/en/h2020-section/science-and-society
Fürst, D., Lahner, M., & Zimmermann, K. (2004). *Neue Ansatze integrierter Stadtteilentwicklung: Placemaking und Local Governance*. Leibnitz-Institut für Regionalentwicklung und Strukturplanung (IRS).
Gibbons, M., Limoges, C., Nowotny, H., Schwartzman, S., Scott, P., & Trow, M. (1994). *The new production of knowledge: The dynamics of science and research in contemporary societies*. Sage.
Haldrup, M., & Bærenholdt, J. O. (2015). Heritage as Performance. In E. Waterton & S. Watson (Eds.), *The palgrave handbook of contemporary heritage research* (pp. 52–68). Palgrave Macmillan.
Hein, C., & van Mil, Y. (2020). Mapping as gap finder: Geddes, Tyrwhitt and the comparative spatial analysis of port city regions. *Urban Planning, 5*(2), 152–166. https://doi.org/10.17645/up.v5i2.2803
Hein, C., van Mil, Y., & Momirski, L. A. (2020). *The waterwheel: A socio-spatial method for understanding and displaying holistic water systems*. Paper presented at the second international conference: Water, megacities and global change.
Hesse-Biber, S. (2010). *Mixed methods research: Merging theory with practice*. Guilford Press.

Hosagrahar, J. (2017). A history of heritage conservation in city planning. In C. Hein (Ed.), *The Routledge handbook of planning history* (pp. 441–455). Routledge.

Jasanoff, S., & Kim, S. H. (2015). *Dreamscapes of modernity: Sociotechnical imaginaries and the fabrication of power*. University of Chicago Press.

Johnson, B. J., & Halegoua, G. R. (2014). Potential and challenges for social media in the neighborhood context. *Journal of Urban Technology, 21*(4), 51–75.

Jonuschat, H. (2012). *The strength of very weak ties – Lokale soziale Netze in Nachbarschaften und im Internet*. Dissertation presented at the HU Berlin. http://edoc.hu-berlin.de/dissertationen/jonuschat-helga-2012-06-12/PDF/jonuschat.pdf

Kunt, G., Székely, J., & Vajda, J. (2017). Making (Dis)Connections: An interplay between material and virtual memories of the Holocaust in Budapest. *Urban People - Lide Mesta, 19*(2), 295–319.

Kvale, S. (2008). *Doing interviews*. Sage.

Lund Declaration. (2009). *New worlds – New solutions. Research and innovation as a basis for developing Europe in a global context*. Conference and declaration, Lund, Sweden, 7–8 July 2009. http://www.eurosfaire.prd.fr/7pc/doc/1247650029_lund_declaration_09_07_2009.pdf

Lund Declaration. (2015). *Lund declaration: Update*. http://jpi-ch.eu/wp-content/uploads/LundDeclaration2015.pdf

Macnaghten, P., & Owen, R. (2011). Good governance for geoengineering. *Nature, 479*, 293.

Manahasa, E., & Manahasa, O. (2020). Defining urban identity in a post-socialist turbulent context: The role of housing typologies and urban layers in Tirana. *Habitat International, 102*, 102202.

Martin, A. M., & Bezemer, P. M. (2020). The concept and planning of public native housing estates in Nairobi/Kenya, 1918–1948. *Planning Perspectives, 35*(4), 609–634. https://doi.org/10.1080/02665433.2019.1602785

Menezes, M. (2019). Digital in action in a neighbourhood in transformation: Notes from Mouraria in Lisbon. In M. Menezes & C. Smaniotto Costa (Eds.), *Neighbourhood & city. Between digital and analogue perspectives* (pp. 25–34). Edições Universitárias Lusófonas. http://cyberparks-project.eu/sites/default/files/publications/kimic_et_al_-_18092-47782-1-pb.pdf

Mieg, H. A. (2015). Social reflection, performed role-conformant and role-discrepant responsibility, and the unity of responsibility: a social psychological perspective. *Soziale Systeme, 19*(2), 259–281.

Mieg, H. A., & Evetts, J. (2018). Professionalism, science, and expert roles: A social perspective. In K. A. Ericsson, R. R. Hoffman, A. Kozbelt, & A. M. Williams (Eds.), *The Cambridge handbook of expertise and expert performance* (2nd ed., pp. 127–148). Cambridge University Press.

Mieg, H. A., & Töpfer, K. (2013). *Institutional and social innovation for sustainable urban development*. Routledge.

Nikšič, M. (2018). Participatory revitalisation of urban public open space: Urban planners' skills needed for improvement of urban public spaces in participatory manner. In N. Novaković, J. P. Grom, & A. Fikfak (Eds.), *Realms of urban design: Mapping sustainability* (pp. 197–214). TU Delft Open.

Nikšič, M., Tominc, B., & Goršič, N. (2018). Revealing residents' shared values through crowdsourced photography: Experimental approach in participatory urban regeneration. *Urbani Izziv, 29*(supplement), 29–42. https://doi.org/10.5379/urbani-izziv-en-2018-29-supplement-002

Nowotny, H., Scott, P., & Gibbons, M. (2001). *Re-thinking science: Knowledge and the public in an age of uncertainty*. Polity Press.

Oevermann, H., & Gantner, E. (Eds.). (2019). *Securing urban heritage: Agents, access, and securitization*. Routledge.

Ren, I. Y., & Bartunek, J. M. (2020). Creating standards for responsible translation of management research for practitioners. In O. Laasch, R. Suddaby, R. E. Freeman, & D. Jamali (Eds.), *Research handbook of responsible management*. Edward Elgar Publishing.

Rifkin, J. (2014). *The zero marginal cost society: The internet of things, the collaborative commons, and the eclipse of capitalism*. St. Martin's Press.

Rynes, S. L., Bartunek, J. M., & Daft, R. L. (2001). Across the great divide: Knowledge creation and transfer between practitioners and academics. *Academy of Management Journal, 44*(2), 340–355.

Scholz, R. W. (2013). Transdisciplinarity. In H. A. Mieg & K. Töpfer (Eds.), *Institutional and social innovation for sustainable urban development* (pp. 305–322). Earthscan.

Shah, A. (2017). Ethnography?: Participant observation, a potentially revolutionary praxis. *HAU: Journal of Ethnographic Theory, 7*(1), 45–59. https://doi.org/10.14318/hau7.1.008

Sharma, G., & Bansal, P. (2020). Cocreating rigorous and relevant knowledge. *Academy of Management Journal, 63*(2), 386–410.

Smith, L. (2006). *Uses of heritage*. Routledge.

Sommer, R. (2007). *Personal space: The behavioral basis of design*. Bosko Books.

Stilgoe, J., Owen, R., & Macnaghten, P. (2013). Developing a framework of responsible innovation. *Research Policy, 42*(9), 1568–1580.

Stoller, P. (1989). *The taste of ethnographic things: The senses in anthropology*. University of Pennsylvania Press.

Stoller, P. (2009). *The power of the between: An anthropological odyssey*. University of Chicago Press.

Strauss, C. (2006). The imaginary. *Anthropological Theory, 6*(3), 322–344.

Sutcliffe, H. (2011). *A report on responsible research and innovation*. MATTER & European Commission. https://ec.europa.eu/programmes/horizon2020/sites/default/files/rri-report-hilary-sutcliffe_en.pdf

Tsoukas, H. (2009a). Creating organizational knowledge dialogically: An outline of a theory. In T. Rickards, M. A. Runco, & S. Moger (Eds.), *The Routledge companion to creativity*. Routledge.

Tsoukas, H. (2009b). A dialogical approach to the creation of new knowledge in organizations. *Organization Science, 20*(6), 941–957.

Tsoukas, H. (2019). *Philosophical organization theory*. Oxford University Press.

Urbact. (2019, October 17). *How participatory placemaking can help URBACT Local Groups to develop urban actions for public spaces in our cities*. https://urbact.eu/how-participatory-placemaking-can-help-urbact-local-groups-develop-urban-actions-public-spaces-our

Van de Ven, A. H. (2007). *Engaged scholarship: A guide for organizational and research knowledge*. Oxford University Press.

Van de Ven, A. H., & Johnson, P. E. (2006). Knowledge for theory and practice. *Academy of Management Review, 31*(4), 802–821.

Von Schomberg, R. (2011). Prospects for technology assessment in a framework of responsible research and innovation. In M. Dusseldorp & R. Beecroft (Eds.), *Technikfolgen abschätzen lehren: Bildungspotenziale transdisziplinärer Methoden* (pp. 1–19). VS Verlag. https://papers.ssrn.com/sol3/papers.cfm?abstract_id=2439112

Von Schomberg, R. (2013). A vision of responsible research and innovation. In R. Owen, J. Bessant, & M. Heintz (Eds.), *Responsible innovation: Managing the responsible emergence of science and innovation in society* (pp. 51–74). Wiley.

Waterton, E., & Watson, S. (Eds.). (2015). *The Palgrave handbook of contemporary heritage research*. Palgrave.